P9-BIP-735

Table of Contents

Copyright © 2006 SingaporeMath.com Inc., Oregon

Using this Guide

This book is a *guide* for teachers using the Primary Mathematics curriculum. It is designed to help teachers understand the course material, to see how each section fits in with the curriculum as a whole, and to prepare the day's lesson. The course material is divided into 70 sessions. Sessions can be extended by including additional problems from supplementary material, or bringing in review from earlier levels of *Primary Mathematics* as needed.

This guide is designed to be used with both the U.S. edition and the third edition of *Primary Mathematics*. U.S. conventions and spellings are used in this guide, such as using commas for thousands and colons for time, and not using "and" in writing out whole numbers in words. However, any items specific to either edition, such as different answers, different page numbers, and different exercise numbers, are indicated.

Workbook exercises can be gone over in class or assigned as homework.

This guide schedules reviews as blocks of several successive sessions. However, you can have a review session every week instead, using selected problems from the textbook and workbook review exercises.

Suggested Material

Game pieces
Two kinds, large and small. Large ones can represent unknowns, and smaller ones units for simplifying algebraic expressions.

Models of solids
Cuboids, prisms, pyramids, cylinders.

Counters or Connect-a-cubes
Use the opaque round counters, cubes, or any suitable counter. They should be in 4-5 different colors.

Optional Resources

Wiggle Woods CD-ROM

This CD-ROM contains learning activities and two games. The name of the program refers to the bug theme. Topics covered include material from both Primary Mathematics 5 and 6. The following chart correlates the different activities to the appropriate part of *Primary Mathematics 6A*.

Copyright © 2006 SingaporeMath.com Inc., Oregon

Primary Mathematics 6A		*Wiggle Woods Primary Six*
Unit 3 – Ratio	Part 1 – Ratio and Fraction	Ratio: Learn and Explore, Activity
	Part 2 – Ratio and Proportion	Ratio: Challenge
Unit 4 - Percentage	Part 2 – One Quantity as a Percentage of Another	Percentage: Learn and Explore, Activity, Challenge Game 2, Level 5
Unit 5 – Speed	Part 1 – Speed and Average Speed	Speed: Learn and Explore, Activity, Challenge

Supplemental Workbooks

These optional workbooks provide a source of extra problems for more practice, tests, and class discussions. Some have interesting and thought-provoking non-routine problems for discussion.

Extra Practice for Primary Mathematics 6 (U.S. Edition)

This workbook has two to four page exercises covering topics from *Primary Mathematics 6A* and *Primary Mathematics 6B*. The level of difficulty and format of the problems is similar to that of the *Primary Mathematics*. Answers are in the back.

Primary Mathematics Challenging Word Problems 6 (U.S. Edition)

This workbook has word problems only. The problems are topically arranged, with the topics following the same sequence as *Primary Mathematics 6A* and *6B*. Each topic starts with three worked examples, followed by practice problems and then challenge problems. Although the computation skills needed to solve the problems is at the same level as the corresponding *Primary Mathematics*, the problem solving techniques necessary in the challenge section are sometimes more advanced, with the problems requiring more steps to solve. It is a good source, though, of extra word problems that can be discussed in class or of enrichment problems for more capable students. Answers are in the back.

Primary Mathematics Intensive Practice 6A (U.S. Edition)

This supplemental workbook has one set of problems for each topic in *Primary Mathematics*. Each topical exercise has questions of varying levels of difficulty, but the difficulty level is usually higher than that in the *Primary Mathematics* textbook or workbook. Some of the word problems are quite challenging and require the students to extend their understanding of the concepts and develop problem solving abilities. There is also a section called "Take the Challenge!" with non-routine problems that can be used to further develop students' problem solving abilities. Answers are located in the back.

Copyright © 2006 SingaporeMath.com Inc., Oregon

This page is blank.

Copyright © 2006 SingaporeMath.com Inc., Oregon

Unit 1 – Algebra

Objectives

- Use letters to represent unknown numbers.
- Write an algebraic expression in one variable.
- Evaluate algebraic expressions in one variable using substitution.
- Simplify algebraic expressions in one variable by adding or subtracting like terms.

Suggested number of sessions: 6

	Objectives	Textbook	Workbook	Activities
Part 1 : Algebraic Expressions				**6 sessions**
1	▪ Use letters to represent unknown numbers. ▪ Write an algebraic expression in one variable using addition or subtraction. ▪ Find the value of an algebraic expression by substitution.	p. 6 p. 7, tasks 1-3		1.1a
2	▪ Write an algebraic expression using multiplication or division. ▪ Find the value of an algebraic expression by substitution.	pp. 8-9, tasks 4-9	Ex. 1	1.1b
3	▪ Write an algebraic expression using more than one operation.	pp. 10-11, tasks 10-14	Ex. 2, #1-2	1.1c
4	▪ Use exponents in algebraic equations. ▪ Evaluate algebraic expressions.	p. 11, tasks 15-17	Ex. 2, #3	1.1d
5	▪ Simplify simple algebraic expressions by using addition or subtraction to combine like terms.	pp. 12-13, tasks 18-21	Ex. 3	1.1e
6	▪ Practice	p. 14, Practice 1A		1.1f

Part 1: Algebraic Expressions	**6 sessions**

Objectives

- Use letters to represent unknown numbers.
- Write simple algebraic expressions with one unknown.
- Find the value of a simple algebraic expression using substitution.
- Write algebraic expressions with one unknown involving more than one operation.
- Evaluate algebraic expressions using substitution.
- Simplify algebraic expressions with one unknown by addition and subtraction of algebraic terms.

Materials

- Game playing pieces to represent variables and unit cubes from a base-10 set or counters to represent constants

Homework

- Workbook Exercise 1
- Workbook Exercise 2
- Workbook Exercise 3

Notes

Arithmetic expressions can involve addition, subtraction, multiplication and division. Now, students will learn about arithmetic expressions which involve an unknown value, and are called *algebraic expressions*. In algebraic expressions, the unknown value is represented by a letter such as n, and we are free to assign any value to the letter. For example, if we *substitute* a value of 3 for n, then the expression $n + 2$ can be *evaluated* as **3** + 2 = 5, and the expression is then equal to 5. If instead we *substitute* a value of 5 for n, the expression can be *evaluated* as **5** + 2 = 7.

At this level, students will only encounter algebraic *expressions*. In later grades students will learn about algebraic *equations* with the answer to the expression given; this will allow us to solve for the unknown. For example, if we are told $n + 2 = 6$, we can find that $n = 4$.

In later grades students will learn that letters which represent an unknown are called *variables*. Variables can be assigned a range of values. Letters as variables can also stand for constants, parameters, or can be used to name objects. The complete definition involved in the concept of *variable* is more complex than students need in pre-algebra, and the term *variable* is not used in *Primary Mathematics*. For this unit, students should think of the letter as standing for a number in an arithmetic expression. The expression can be evaluated by assigning a specific value to the letter. The letter then becomes that number.

Students have been exposed to the idea of an unknown value starting in *Primary Mathematics 1A*. The unknown was the missing part in a number bond, or an empty rectangle, ☐ + 4 = 34, or a blank line, 43 + ____ = 100.

Copyright © 2006 SingaporeMath.com Inc., Oregon

In *Primary Mathematics 3A*, students were introduced to diagrams in which an unknown quantity was represented by a bar labeled with "?". Now, we will be using a letter to represent the unknown value.

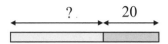

You can help students see the analogy between a letter representing an unknown and the familiar unknown in a bar diagram by labeling the bar whose value is presently unknown with a letter. Since we may not always know the relative size of the unknown bar to bars for known values, you may want to use a dotted line to indicate that the bar could be relatively shorter or longer.

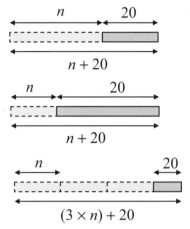

Students should also be familiar with diagrams using unit bars of unknown value. These could be labeled with a letter representing the unknown value, rather than with "?". Here, the 3 unit bars could be longer or shorter than 20, depending on the value of *n*.

In algebraic equations involving whole numbers, the expression can also be visualized as bags or boxes holding an unknown number of unit cubes or other items. Each bag represents the same letter and each holds the same number of items.

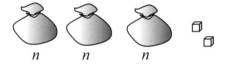

Total: $(3 \times n) + 2$
or $3n + 2$

Note that when expressing the multiple of a letter, the custom is to remove the multiplication sign, so rather than $3 \times n$, we write $3n$. The unknown is written second, so we write $3n$ rather than $n3$ (even though multiplication is commutative). When there is only 1 of *n*, the custom is to just write "*n*".)

Exponents are used when a number is multiplied by itself one or more times, as: $3 \times 3 = 3^2$, or $3 \times 3 \times 3 \times 3 = 3^4$. In the same way, a letter multiplied by itself one or more times is also expressed with an exponent. For example, $n \times n \times n \times n = n^4$. And, $3n^4 = 3 \times n \times n \times n \times n$.

Algebraic expressions involving division are normally written as fractions. For example, $m \div 3$ is usually represented as $\frac{m}{3}$. During discussion, you may want to remind students that just as $12 \div 3$ is the same as $\frac{12}{3}$, so $m \div 3$ is the same as $\frac{m}{3}$.

In algebraic expressions, algebraic *terms* are separated by + or – signs. The expression $2y + 3$ has two terms, while $3y + 4y - 3$ has 3 terms, $3y$, $4y$, and 3 (or – 3).

In a term containing both a number part and an unknown part, such as $2y$, the number part of the term, 2, is called a *coefficient*. In $5y$, the coefficient is 5. A term that contains only numbers (no unknowns) is called a *constant*. In $2y + 3$, the 3 is a constant.

Copyright © 2006 SingaporeMath.com Inc., Oregon

Like terms are those terms in an expression that have the exact same letters *and* exponents. $2y$ and $3y$ are like terms, $2a$ and $2b$ are not, and $2y$ and $3y^2$ are not.

Addition and subtraction can be done in any order, as long as the plus or minus signs are kept with the number (otherwise, addition and subtraction must be done from left to right). In performing an arithmetic operation on expressions, we are simplifying the expression. For example:

$$3 - 10 + 8 = 3 + 8 - 10 = 1$$

Similarly, algebraic expressions can also be simplified by combining like terms and constants. We can do this by first grouping like terms. For example:

$$3y - 10y + 8y = 3y + 8y - 10y = y$$

or

$$4y - 5 - 2y + 10 = 4y - 2y + 10 - 5 = 2y + 5$$

The sign before each term must stay with its term. The fact that the "–5" term can be moved after the "+10" term can cause confusion for some students. Students have learned that addition is commutative and can be done in any order, but that subtraction is not. In *Primary Mathematics 5A*, students learned to add and subtract from left to right. This automatically kept the minus sign with the term that follows it. Until students learn about negative numbers, there will be occasions when they will need to rearrange the terms in an expression to avoid negative numbers. If so, they need to remember to keep the minus sign with the term it applies to.

In this unit, students will only encounter expressions with like terms in one unknown and constants; that is, they will not encounter expressions such as $3y + 4x - y + 10 + 7y$, which has two unknowns, x and y.

Illustrate combining like terms concretely. You can draw bags and marbles as in the text to illustrate the problems. Each bag contains the same unknown number of marbles, and the marbles each stand for a one. For $4y - 5 - 2y + 10$, you could draw 4 bags for $4y$ and 10 marbles for 10, and then take away two of the bags and 5 marbles by crossing them out. The first term and the ones with "+" in front indicate what is added in, and the terms with "–" in front indicate what is taken out. Add in everything first, then take out. Make sure students understand the bags in any one problem all hold the same number of marbles, but in another problem we can have them holding a different number of marbles, even though the letter used for the unknown is the same as that in the earlier problem.

Students can also use manipulatives to help them understand the process. Choose two types of uniform objects, one object to represent the unknown, and another (smaller) object to represent Ones (constants). For example, a playing piece or a marker from a game could represent the unknown. Unit cubes from a base-10 set can represent the constants. Students can set out markers and cubes for terms to be added, then take away some for terms to be subtracted. For example, for $4y - 5 - 2y + 10$, they can set out 4 markers for $4y$ and 10 cubes for 10, and then take away two of the markers and 5 cubes. The first term and the ones with "+" in front indicate what are added in first, and the terms with "–" in front indicate what is then taken out.

Copyright © 2006 SingaporeMath.com Inc., Oregon

Activity 1.1a **Algebraic expressions with addition or subtraction**

1. Discuss using letters to represent an unknown number
   - Discuss the top half of **p. 6 in the textbook**.
     - Lead students to see that Limei is 2 years older than Angela.
   - Draw the table for Limei's and Angela's ages on the board and write an equation in the second column.

Angela's age	Limei's age
6	$6 + 2 = 8$
7	$7 + 2 = 9$
8	$8 + 2 = 10$
9	$9 + 2 = 11$
10	$10 + 2 = 12$
n	$n + 2$

   - Ask students if they see a pattern in the equations. In each, 2 years are added to Angela's age.
     - Tell students that if we don't know Angela's age at first, or if we want a general expression to show how Limei's age relates to Angela's age, we can use the letter n to stand for Angela's age. The letter n can be any number that makes sense within the context of this situation and represents Angela's age.
     - Limei's age could then be written as $n + 2$. Add another line to the table to show this.
   - Point out that we are *expressing* Limei's age *in terms of* Angela's age. If we say that n stands for Angela's age, then we can say that we are expressing Limei's age in terms of n.
   - If we are told what Angela's age is, we can *substitute* the number we are given for Angela's age for n, and then find Limei's age.
   - Ask students: how old will Limei be when Angela is 56?

   Limei's age for $n = 56$:
   $n + 2 = 56 + 2 = 58$

     - Write the algebraic expression and substitute 56 for n.
   - Tell students that $n + 2$ is called an *algebraic expression*. It contains an unknown number represented by n. When the value of n changes, the value of the expression $n + 2$ changes accordingly.
   - Show students how the expression can be represented by a bar diagram. Since we don't know the bar length for n relative to the bar length for 2 until we assign a number to n, we can think of n's bar length as being elastic — it can stretch or shrink depending on the value for n.

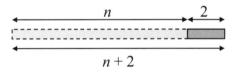

2. Build algebraic expressions which involve addition or subtraction.
   - Draw two items on the board, such as a book and a notepad.
     - Tell students that the book costs \$4 and the notepad costs \$3 and write those values on the board. Ask students to write an equation and find their total cost.

book	notepad	total
\$4	\$3	$\$4 + \$3 = \$7$
\$4	\$$n$	$\$(4 + n)$
		$\$(n + 4)$
\$$b$	\$3	$\$(b + 3)$

     - On the next line, replace the 3 for notebook with n. Ask students: if instead we write the notebook cost as \$$n$, how do we write an expression for the total cost? Write $\$(4 + n)$.

Copyright © 2006 SingaporeMath.com Inc., Oregon

- o Under the $(4 + n)$, write $(n + 4)$. Ask students if this is the same thing. It is, since addition can be done in any order. Tell students that in expressions which involve adding or subtracting a number and an unknown, when they can, they should write the part of the expression with the letter for the unknown first, since this is the convention that will be used later in algebra.
  - o On the next line, replace the $4 for the book with b. Ask students to write an expression for the total cost.

- Similarly, have students write an equation for the following or a similar situation:
  - o John had 35 marbles. He lost 8 of them. How many did he have left? $35 - 8 = 27$
  - o John had 35 marbles. He lost x of them. How many did he have left? $35 - x$
  - o John had m marbles. He lost 8 of them. How many did he have left? $m - 8$

- Discuss **tasks 1-2, textbook p. 7**. Note that in 1(b) the terms are reversed in order to write the x first, rather than writing $8 + x$.

3. Use substitution to find the value of an algebraic expression.
   - Discuss **task 3, textbook p. 7**.
   - You may want to illustrate task 3 with a part-whole model.

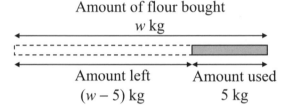

Activity 1.1b   **Algebraic expressions with multiplication or division**

1. Build algebraic expressions which involve multiplication or division.
   - Discuss **task 4, textbook p. 8**.
     - o Lead students to see the pattern — the second factor in the equation is the same as the number of packages. So if we let n equal the number of packages, we can write a general expression $4 \times n$ for the total number of apples.

2. Use this chance to discuss some customary usage for algebraic expression and some vocabulary.
   - Point out that in algebraic expressions, we can omit the multiplication sign and write $4 \times n$ as $4n$. We also write $n \times 4$ as $4n$ since multiplication can be done in any order.
   - Tell students that in $4n$, the "4" part is called the *coefficient*, and "$4n$" is called a *term*. A term is an unknown and the coefficient it is multiplied by. When the coefficient is 1, the term is $1n$, but the custom is to write $1n$ as simply n.
   - In a term, we always write the coefficient first. So we write $4n$, even though this might represent an unknown number of 4's (i.e., n fours) rather than 4 times an unknown number (i.e. 4 n's). Whatever we substitute for n, $4n$ is the same as $n4$, but by convention, we will always write $4n$.
   - A constant is also called a term. So in the expression $n + 4$, there are two terms, n and 4.

Copyright © 2006 SingaporeMath.com Inc., Oregon

- In an expression with two terms, such as one where we are adding 4 and *n*, we normally write the term with the unknown first, *n* + 4, but that isn't required yet, and 4 + *n* is acceptable. In 4 − *n* we also have two terms, but for now we write the term with the unknown last.
- Tell students, suppose we have three crates, each crate containing n bags, each bag containing 4 apples. We can write the total number of apples as 3 × 4*n*. In this expression, we cannot eliminate the multiplication symbol between the two numbers, but we can between the coefficient and the unknown in the second term.
- In some textbooks, the multiplication sign is often replaced by a raised dot, with 3 × 4*n* shown as 3 • 4*n*. In *Primary Mathematics* a dot is not used, but students should be told that they may see algebraic expressions elsewhere with a raised dot • for the "×" symbol.
- The custom in most texts, including *Primary Mathematics*, is to italicize the letter being used for an unknown. If there is likely to be any confusion, between × (times) and *x* (unknown) don't use the letter *x* for the unknown.

3. Guide students in writing algebraic expressions for a situation that involves multiplication.
   - Write a simple word problem and use boxes for the numbers. First fill the boxes with suitable numbers, and then replace one of the numbers with a letter. For example:
     - Notebooks cost $ ☐ each. How much do ☐ notebooks cost?
     - Notebooks cost $ 3 each. How much do 6 notebooks cost? $3 × 6 = $18

       Draw a part-whole model to show the multiplication concept.

       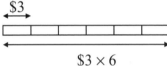

     - Notebooks cost $ *n* each. How much do 6 notebooks cost? $*n* × 6 = $6*n*

       Point out that we write 6*n* rather than *n*6.

       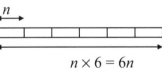

     - Notebooks cost $ each. How much do *z* notebooks cost? $3 × *z* = $3*z*
     - Lead students to see that when *z* = 6, then 3*z* = 3 × 6 = 18. Ask students to evaluate 3*z* if *z* = 10; if *z* = 15.

   - Guide students in writing algebraic expressions for a situation that involves division. For example:
     - Marcus has ☐ music CDs. He divides them into ☐ equal groups. How many CDs are in each group?
     - Marcus has 24 music CDs. He divides them into 3 equal groups. How many CDs are there in each group? 24 ÷ 3 = 8

       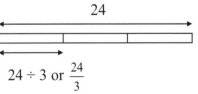

     - Remind students that we can write division problems as fractions. $24 \div 3 = \frac{24}{3} = 8$

Copyright © 2006 SingaporeMath.com Inc., Oregon

o Marcus has $\boxed{m}$ music CDs. He divides them into $\boxed{3}$ equal sets. How many CDs are in each group? $\dfrac{m}{3}$

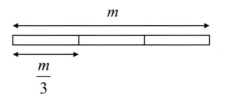

- Tell students to write $m \div 3$ as $\dfrac{m}{3}$. We write algebraic expressions involving division as fractions. Lead students to see that if $m = 24$, then $\dfrac{m}{3} = 8$

o Ask students how many CDs would be in each group if Marcus had 30 CDs.

- Point out that $\dfrac{m}{3} = \dfrac{1}{3}m$. $\dfrac{1}{3}$ is the coefficient in the equivalent terms $\dfrac{1}{3}m$ or $\dfrac{m}{3}$.

4. Discuss **tasks 5-8, textbook pp. 8-9**.

5. Have students do **task 9, textbook p. 9**.

6. Ask students to make up word problems to go with the algebraic expressions in task 9. For example, for $n + 4$: There are n boys. There are 4 more girls than boys. How many girls are there?

Workbook Exercise 1

Activity 1.1c   **Algebraic expressions with all four operations**

1. Build algebraic expressions which involve all four operations.
   - Refer to **task 10, textbook p. 10**.
     o Draw 5 bags on the board, and label each one with x.
     o Tell students there are x marbles in each bag.
     o Ask them how many marbles there are in all ($5x$).
     o Draw 3 marbles next to the bags and ask how many marbles there are now ($5x + 3$).
     o Then, ask how many marbles there would be if $x = 10$. Write the equation on the board.
   - Point out that in evaluating the expression, we follow order of operations and do the multiplication first.
   - Review order of operations if necessary. Order of operations was taught in unit 1 of *Primary Mathematics 5A*. In an expression involving several terms and operations, we do multiplication and division from left to right, then addition and subtraction.
   - Give students the following word problems and lead them to tell you all the different ways to write expressions to represent the given information:
     o Sam had 5 bags of marbles, with 10 marbles in each bag. He added 3 marbles to each bag. How many marbles does he now have? $(10 + 3) \times 5$ or $5 \times (10 + 3)$ or $5 \times 10 + 5 \times 3$. For the first two expressions, we had to use parentheses to show that we must add the extra marbles before we multiply by the number of bags.

$(10 + 3) \times 5$
or $5 \times (10 + 3)$
or $5 \times 10 + 5 \times 3$

Copyright © 2006 SingaporeMath.com Inc., Oregon

- o Sam had 5 bags of marbles, with *m* marbles in each bag. He added 3 marbles to each bag. How many marbles does he now have? $(m + 3) \times 5$ or $5 \times (m + 3)$ or $5m + 15$.

$(m + 3) \times 5$
or $5 \times (m + 3)$
or $5m + 5 \times 3$
$= 5m + 15$

- • Discuss **task 11, textbook p. 10**.
  - o Tell students that we call expressions like $(5x + 3)$ *algebraic expressions*.
  - o We describe algebraic expressions as having — or being made up of — terms. For example, the algebraic expression $(5x + 3)$ has two terms: one is the $5x$, the other is the 3.
  - o Have students tell you how many terms the expression $(5x + 3 - 2x - 1)$ has.
  - o Write on the board some other expressions with different numbers of terms, some algebraic and some numerical, and have students tell you how many terms each expression has.

- • Discuss **task 12, textbook p. 10**.
  - o You can ask students to draw a bar model to illustrate this problem.
  - o For (a), point out that since we don't know how large *y* is in relation to 50, we can't make the drawing proportional. But a drawing can still help us come up with an appropriate algebraic expression, $(50 - y) \div 2$.

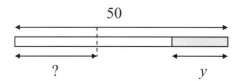

  - o For (b), note that by writing the expression as a fraction, we see right away that we have to subtract first before dividing. $\dfrac{50 - y}{2} = (50 - y) \div 2$. Emphasize that it would not be correct to write the expression $50 - y \div 2$ for this problem.
  - o Ask students how much money each daughter would receive if *y* were a different number, such as $16.
  - o If a student asks about the number of terms in this expression, you can tell them there are two in the numerator and one in the denominator.

2. Provide some additional examples, such as the following. (Do not use problems that would result in an unknown in the denominator.)

Ann's allowance is $*m* each week. How much allowance will she be given in 4 weeks? (In 4 weeks she will get $4*m*.)

During those 4 weeks, Ann also earned $12 from babysitting. How much did she get in those 4 weeks, altogether? $(4*m* + 12)

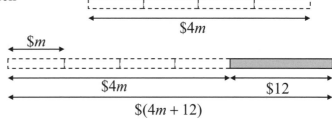

How much money would she have made if her allowance were $12 a week? ($4 \times \$12 + \$12 = \$60$)

What would she have made if her allowance were $5 a week? ($4 \times \$5 + \$12 = \$32$)

Copyright © 2006 SingaporeMath.com Inc., Oregon

During 4 weeks, Paula made $100. b of it was from babysitting during that time, and the rest from a weekly allowance. Write an expression for Paula's weekly allowance. $\left(\dfrac{\$100-b}{4}\right)$

What would her allowance be if she made $40 from babysitting? $\left(\dfrac{\$100-40}{4}=\$15\right)$

3. Discuss **tasks 13-14, textbook pp. 10-11**.
   - Point out that in evaluating the algebraic expression for 14(b) we have to follow order of operations. Remind students that $45 - 3r$ is $45 - 3 \times r$. Following order of operations, we multiply $3 \times r$ first, and then subtract that from 45. After we evaluate the numerator, we can then divide the result by 3.

4. Ask students to make up word problems that would fit the algebraic expressions given in tasks 13 and 14.

   Workbook Exercise 2, problems 1-2

Activity 1.1d   **Exponents**

1. Build algebraic expressions which involve exponents.
   - Tell student you have a cube that measures 4 cm on the side. Ask them to write an equation for its volume. $4 \text{ cm} \times 4 \text{ cm} \times 4 \text{ cm} = 64 \text{ cm}^3$
   - Now, ask them to write an expression for the volume if the side of the cube measures s cm. The volume is s cm $\times$ s cm $\times$ s cm.
   - Tell students that customarily we don't write this $s \times s \times s$ cm^3 or as sss cm^3, instead we write it as s^3 cm^3. The ^{3} in this expression is called an *exponent*.
   - Emphasize that the correct notation is to write the exponent both for the letter (s) and the unit (cm).
   - Now, tell students that you have a box with a square base, 4 cm on the side. Ask: what is the area of the base? It is $4 \text{ cm} \times 4 \text{ cm} = 16 \text{ cm}^2$. Ask them to write an equation for the volume of the box, if its height is 5 cm. $4 \text{ cm} \times 4 \text{ cm} \times 5 \text{ cm} = 80 \text{ cm}^3$.
   - Now, ask them to write an expression for the area of the base of the box if the side of the base is b cm. It is $b \text{ cm} \times b \text{ cm} = b^2 \text{ cm}^2$. Then ask them to write an equation for the volume of this box, if its height is 5 cm. The volume is $b \text{ cm} \times b \text{ cm} \times 5 \text{ cm} = 5b^2 \text{ cm}^3$. Point out that since we are finding volume, the volume is centimeter cubed, and we write an exponent of 3 for the cm. But the exponent of the unknown (b) in the expression is 2, not 3, because we are only multiplying 2 b's together.
   - Ask your student to write out some expressions where the unknown has an exponent using multiplication. For example:
     $m^4 = m \times m \times m \times m$
     $m^4 - 2 = m \times m \times m \times m - 2$
   - Tell students to be careful **not** to evaluate an expression such as m^4 as $m \times 4$. Ask them to evaluate $m^4 - 2$ and $4m - 2$ for $m = 3$.

2. Discuss tasks **15-16, textbook p. 11**.

Copyright © 2006 SingaporeMath.com Inc., Oregon

3. Have students do **task 17, textbook p. 11**.

   Workbook Exercise 2, problem 3

Activity 1.1e   **Simplify algebraic expressions**

1. Illustrate simplifying simple algebraic expressions.
   - Discuss **task 18, textbook p. 12**.
     - In (a), we start out with 4 bags, each containing x beads, and add 3 more bags. We now have 7 bags, each holding x beads, or $7x$ beads total. So, we can simplify $4x + 3x$, using addition.
     - Draw a part-whole model. Label the unknown unit as x. So x is a unit, and we already know that 4 units + 3 units = 7 units. So $4x + 3x = 7x$.
     - For (b), draw a comparison model; one bar with 4 units, and one with 3 units. The difference between the bars is 1 unit. Since each unit represents x beads, the difference is $1x$, or x beads. We do not have to write 1 in front of the letter x, since $1x = x$.

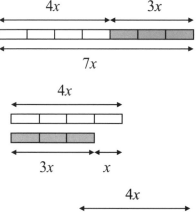

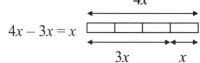

$4x - 3x = x$

   - Show students a similar problem which involves subtraction as take-away rather than as comparison. Tell students that we start with 4 bags of red beads, and take away 3 bags. How many beads are left? This time, we draw a part-whole diagram. The algebraic expression is the same, and we are left with 1 unit, x beads.

2. Illustrate simplifying algebraic expressions with more terms.
   - Draw 6 bags of n marbles and write $6n$.
   - Cross out 3 bags, and ask students how we would write an expression to show what we did. $6n - 3n$.
   - Draw 2 more bags, and have students change the expression to show this. Ask, how many marbles we have now? $6n - 3n + 2n = 5n$.
   - Point out that we are simplifying the expression $6n - 3n + 2n$ by *combining* the like terms to get $5n$ in the same way that $6 - 3 + 2 = 5$.
   - Now, ask students to find the total number of marbles at each step if $n = 10$. We started with $6n$, or 60 marbles, took away $3n$, or 30 marbles $(60 - 30)$ to get 30 marbles. Then we added $2n$, or 20 marbles, to end up with 50 marbles. $(5n = 50$ when $n = 10)$

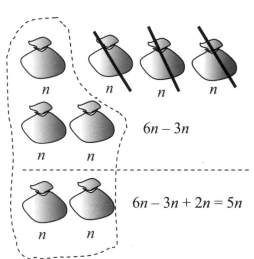

Copyright © 2006 SingaporeMath.com Inc., Oregon

- Write the expression $6n + 2n – 3n$ and illustrate it by drawing 6 bags of n marbles, then two more, then crossing out 3 bags.
- Ask students to simplify the expression. The answer is $5n$, the same as for $6n – 3n + 2n$. Point out that when we simplify the expression, we can take away $3n$ first and then add $2n$, or add $2n$ first and then take away $3n$. So we can add and subtract in any order, so long as we keep the + or – with whatever we are adding in or taking away.

$6n + 2n – 3n = 5n$

3. Illustrate simplifying algebraic expressions which contain constants.
   - Draw 2 bags containing n marbles, and 3 single marbles.
     - Ask students to write an expression for the total number of marbles ($2n + 3$).
     - Add 2 more marbles, add "+ 2" to the expression ($2n + 3 + 2$), and ask students to simplify it ($2n + 5$).
     - Now add another bag of n marbles. Add n to the expression, ($2n + 3 + 2 + n$) and ask students to simplify it ($3n + 5$).

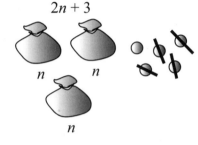

$2n + 3$

     - Cross out 4 marbles, write "– 4" at the end of the expression and ask students to simplify it.
     - Point out to students that terms containing the same letter (unknown) are called *like terms*. They represent a certain number of units. Terms that are just numbers (without an unknown) are also *like terms* to each other, but are different from terms that have a letter. When we simplify expressions, we combine like terms.

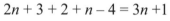

n

$2n + 3 + 2 + n – 4 = 3n + 1$

$2n + n = 3n$
$3 + 2 – 4 = 1$

4. Discuss **tasks 19-20, textbook p. 13**.

5. Have students do **task 21, textbook p. 13**.
   - You can provide students with game pieces and unit cubes to work out the answers. Each game piece represents the unknown value.

6. Optional.
   - Have students simplify the expression $3n – 2 + 1$. The correct answer is $3n – 1$. Some students may give $3n – 3$ as the answer.
   - Ask students how they could visualize $3n – 2$. They could think of 3 bags of n marbles, with two marbles taken out of the third bag. Even though they don't know how many marbles there are in the bags, they do know that two of them have the same number of marbles, and one of them, the one from which two were removed, has 2 fewer marbles than the others.
   - Now, we add one marble back in. We can write that as $3n – 2 + 1$. That gives us 2 bags with n marbles, and a third bag which now it has one less than the others. So we have $3n – 1$ marbles.

Workbook Exercise 3

Copyright © 2006 SingaporeMath.com Inc., Oregon

Activity 1.1f   **Practice**

1. Have students do Practice **1A, textbook p. 14** and share their solutions.
    You can have students draw models for some of the word problems. For example:

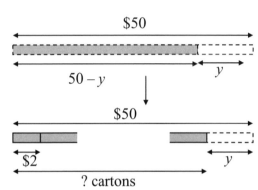

#8. (a) Peter's age = $3x + 4$ years

#9. (a) Number of cartons of milk = $\dfrac{50 - y}{2}$

Huili spent $\$(50 - y)$ for milk, at $2 per carton. So one unit is $2, and we need to find out how many units he bought. To illustrate this, we make $2 units, but we don't know how many units there are. It would not be accurate to show this by dividing the $50 - y$ bar into two equal parts.

2. Optional. Build algebraic equations for number tricks to show how the tricks work.
    *   Ask students to do the following number trick.
        Choose a Number. (To make this easier for students, you can specify a range, such as under 100).
        Add 5.
        Double the result.
        Subtract 4.
        Divide the result by 2.
        Subtract the number you started with.
        The result will be 3
    *   Have students pick other starting numbers and see that the result will always be 3.
    *   Guide students through writing algebraic expressions for each step, using a variable for the number picked:

Pick a number	n
Add 5	$n + 5$
Double the result	$2n + 10$ (note that both terms must be doubled)
Subtract 4	$2n + 10 - 4 = 2n + 6$
Divide the result by 2	$n + 3$ (both terms must be halved)
Subtract the number you started with	$n + 3 - n = 3$ The result is always 3.

Copyright © 2006 SingaporeMath.com Inc., Oregon

- Have students do another number trick, determine the result for several whole numbers, and then prove that the trick works for all whole numbers.

      Pick a number below 1000
      Add 3
      Double the result
      Subtract 4 from the result
      Divide the result by 2
      Subtract the original number

- Students can create their own number tricks.

- Enrichment: This number trick involves two different unknowns and the place value concept:

Pick a number between 0 and 9	n
Double it	$2n$
Add 5	$2n + 5$
Multiply by 5	$10n + 25$
Pick another number between 1 and 9 and add its value to the total	$10n + 25 + m$
Subtract 25 from the total	$10n + 25 + m - 25 = 10n + m$
The result is a 2-digit number where the first digit is the same as the first number picked, and the second digit is the same as the second number picked.	$10n + m$ gives a 2 digit number with the first digit the tens and the second digit the ones.

3. Optional: Use algebraic expressions to prove divisibility by 3.

   In *Primary Mathematics 3A*, students learned divisibility rules for two-digit numbers. They found that a number is divisible by 3 when the sum of the digits is divisible by 3. This was done by having the students check the sum of the digits for numbers from 10 to 99, and relating the result to divisibility by 3. Now, you can show students a proof for this rule using algebraic expressions and extend it to larger whole numbers. This discussion will involve expressions with more than one unknown.

   - Show students that if we add together two or more numbers each of which is divisible by 3, then the sum of these two numbers is also divisible by 3:

   If a is divisible by 3, then $\dfrac{a}{3}$ is a whole number. If b is divisible by 3, then $\dfrac{b}{3}$ is a whole number.

   The sum of two whole numbers is a whole number, so $\dfrac{a}{3} + \dfrac{b}{3}$ is a whole number.

   If $\dfrac{a}{3} + \dfrac{b}{3}$ is a whole number, and $\dfrac{a}{3} + \dfrac{b}{3} = \dfrac{a+b}{3}$, then $\dfrac{a+b}{3}$ is a whole number, and $a + b$ must be divisible by 3.

   So if a is divisible by 3, and b is divisible by 3, then $a + b$ is divisible by 3

   Example: $12 + 6 = 18$. If we know that 12 and 6 are each divisible by 3, then we know that their sum, 18, is also divisible by 3.

Copyright © 2006 SingaporeMath.com Inc., Oregon

- Remind students that the two-digit number 57 can be written as $5 \times 10 + 7$. Substitute b for the number of tens, and a for the number of ones, and show students that all two-digit numbers can be written as $10b + a$.

$57 = 5 \times 10 + 7$
If $b = 5$ and $a = 7$,
then $57 = b \times 10 + a = 10b + a$
Any two digit number can be written as $10b + a$, where
b = number of tens
a = number of ones
$9b + b + a = 10b + a$

- Remind students that $9b$ is divisible by 3 ($9b \div 3 = 3b$). Now, ask student to simplify the expression $9b + b + a$. It is the same as $10b + a$. Any two-digit number can be written as $9b + b + a$.

Since b is the digit in the tens place, and a is the digit in the ones place, then if the sum of b and a is divisible by 3, the whole expression $9b + b + a$ is divisible by 3. So, if the sum of the digits of a two-digit number is divisible by 3, then the whole number is divisible by 3.

- Use 57 as an example.

$5 + 7 = 12$, which is divisible by 3.
So 57 is divisible by 3.
$57 = (9 \times 5) + 5 + 7$

- Show how this can be extended to a 3-digit number.

Any 3-digit number can be written as
$100c + 10b + a$, where
c = number of hundreds
b = number of tens
a = number of ones
$100c + 10b + a = 99c + c + 9b + b + a$
$\qquad\qquad\qquad\quad = 99c + 9b + c + b + a$
Example : $324 = (99 \times 3) + 3 + (9 \times 2) + 2 + 4$
Since $99c$ and $9b$ are divisible by 3, all we have to show is that
if $c + b + a$ is divisible by 3, then
$100c + 10b + a$ is divisible by 3.

- Ask students to check whether 576 is divisible by 3, using three methods: division, the argument given above, and the result proven above.

$576 \div 3 = 192$
$5 + 7 + 6 = 18$, which is divisible by 3, so 576 is divisible by 3
$576 = (100 \times 5) + (10 \times 7) + 6$
$\qquad = (99 \times 5) + 5 + (9 \times 7) + 5 + 7 + 6$
$\qquad = (99 \times 5) + (9 \times 7) + 18$

Copyright © 2006 SingaporeMath.com Inc., Oregon

Unit 2 – Solid Figures

Objectives

- Associate two-dimensional drawings with three-dimensional shapes.
- Visualize pyramids, prisms, and cylinders from two-dimensional drawings.
- Identify nets of cubes, cuboids, prisms, and pyramids.
- Identify the solid represented by a net.
- Determine whether a figure can be the net of a given solid.
- Determine whether a solid can be formed from a given net.

Suggested number of sessions: 5

	Objectives	Textbook	Workbook	Activities
Part 1 : Drawing Solid Figures				**1 session**
7	▪ Associate two-dimensional drawings with three-dimensional solids. ▪ Visualize prisms, pyramids, and cylinders from two-dimensional drawings. ▪ Determine the number and shapes of the faces of a two-dimensional drawing of a solid.	p. 15 p. 16, tasks 1-3	Ex. 4	2.1a
Part 2 : Nets				**4 sessions**
8	▪ Form solids from nets.	p. 17 p. 18, task 1	Ex. 5	2.2a
9	▪ Identify nets of cubes.			2.2b
10	▪ Identify nets of cuboids, pyramids, and prisms.	p. 19, task 2	Ex. 6	2.2c
11	▪ Identify the solid represented by a net.	p. 20, task 3	Ex. 7	2.2d

Copyright © 2006 SingaporeMath.com Inc., Oregon

Part 1: Drawing Solid Figures	1 session

Objectives

- Associate two-dimensional drawings with three-dimensional models or solids.
- Visualize pyramids, prisms, and cylinders from two-dimensional drawings.
- Determine the number and shapes of the faces of a two-dimensional drawing of a solid.

Materials

- Models of solids – cuboids, prisms, pyramids, cylinders

Homework

- Workbook Exercise 4

Notes

In *Primary Mathematics 4*, students learned to relate two-dimensional drawings of cubes and cuboids (rectangular prisms) to solids and to draw them using dot-paper. In this unit, students will learn to recognize and draw the two-dimensional representations of prisms, pyramids, and cylinders.

The Primary Mathematics curriculum does not require students to use the terms prism or pyramid at this time. They should, however, understand the difference between a prism and pyramid. If your students need to know these terms for your state's standards, though, you can teach them during this unit.

A *polygon* is a closed plane figure, with straight sides. Rectangles, squares, and triangles are all polygons.

A *regular polygon* is a polygon in which the sides are all the same length and are symmetrically placed about a common center (i.e., the polygon is both equiangular and equilateral). A square and an equilateral triangle are regular polygons.

A *polyhedron* is a three-dimensional solid which consists of a collection of polygons, joined at their edges.

A general *prism* is a polyhedron possessing two congruent polygonal faces and with all remaining faces parallelograms.

right rectangular prism (cuboid)

right triangular prism

A *right prism* is a prism in which the top and bottom polygons lie on top of each other so that the vertical polygons connecting their sides are not only parallelograms, but also rectangles. If the upper and

Copyright © 2006 SingaporeMath.com Inc., Oregon

lower polygons of a right prism are rectangles, then the prism is a cuboid (also called a rectangular prism, and a rectangular parallelepiped).

A *pyramid* is a polyhedron where one face (known as the "base") is a polygon and all the other faces are triangles meeting at a common polygon vertex (known as the "apex"). A right pyramid is a pyramid for which the line joining the centroid of the base and the apex is perpendicular to the base. A regular pyramid is a right pyramid whose base is a regular polygon.

rectangular   triangular
pyramid   pyramid

All prisms and pyramids in this unit will be right prisms or pyramids.

A *cylinder*, as used here, is a solid bounded by two congruent circular bases, one directly above the other.

cylinder

Copyright © 2006 SingaporeMath.com Inc., Oregon

Activity 2.1a   **2-dimensional drawings**

1. Examine three-dimensional models of cubes, cuboids, prisms, and pyramids.
   - Provide students with three-dimensional models of cubes, cuboids, prisms, pyramids, and cylinders.
   - Have student examine the faces, edges, and curved surfaces of the models. Get them to count the faces and name the shapes of each face.
   - Make sure the students know the following:
     - ➢ A *face* is a flat surface.
     - ➢ An *edge* is a line formed by two faces.
     - ➢ A *vertex* is the point where two or more edges meet.
     - ➢ Prisms have two faces opposite each other (*bases*) that are the same shape and size.
     - ➢ A pyramid has only one base, which may or may not be a triangle. All the other faces are triangles, and meet at a vertex.
     - ➢ The faces of prisms and pyramids are polygons.
   - Similarly, discuss faces and edges or curved surfaces of objects (e.g. tin cans, boxes, prisms for breaking up light waves) or pictures of objects.

2. Relate 3-dimensional models to 2-dimensional drawings.
   - Discuss **page 15 in the textbook**. Discuss the number of faces and shape of the faces.
     - o Get students to tell you which of these figures are prisms, and which are pyramids.
   - Provide students with the 3-dimensional models of prisms and pyramids and have students try drawing a 2-dimensional representation, using the figures on p. 15 to guide them.
   - Discuss **tasks 1-3, textbook p. 16**.
     - o For task 3, guide students to see that A, B, and C are prisms, but D is not.

Workbook Exercise 4

Copyright © 2006 SingaporeMath.com Inc., Oregon

Part 2: Nets	4 sessions

Objectives

- Determine whether a figure can be the net of a given solid.
- Determine whether a solid can be formed from a given net.

Materials

- Paper, scissors, tape
- Square graph paper

Homework

- Workbook Exercise 5
- Workbook Exercise 6
- Workbook Exercise 7

Notes

A net of a solid is a figure which can be folded to form the surface of the solid. A solid can have more than one net.

Being able to visualize the net of a solid will help a student find the surface area of solids in later levels.

Some students have good spatial visualization and can easily identify the net of a solid. Others have more difficulty. If a student has difficulty, have him or her trace the nets in the learning tasks and workbook exercises and try folding each of them into a solid to see concretely why they may or may not form the solid.

Even without tracing and cutting out the shapes, some choices can be easy to rule out. The nets need to have the same number and type of shapes as the faces of the solid. Any point on the net where three or four lines come together is going to be the vertex of the solid. Since adjacent edges of the nets will go together to form an edge, they must be the same length.

Copyright © 2006 SingaporeMath.com Inc., Oregon

Copyright © 2006 SingaporeMath.com Inc., Oregon

Activity 2.2a   **Nets**

1. Create nets from solids.
   - Have students do the activity on **p. 17 of the textbook**.
   - Tell students that the flat figure is called a *net of a solid*. The figure on p. 17 is a net of a cuboid.
   - After students have formed the cuboid, have them tape all the edges. Then have them cut along several other edges to form a different net. Students can compare their nets. They should realize that a solid can have different nets.
   - Have students do **task 1, textbook p. 18**.
   - Get them to draw 2-dimensional representations of the solids.

   (a)   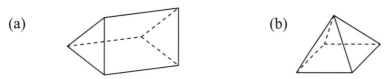   (b)

   - Have students tape up the edges and then see if they can form a different net by cutting along the non-taped edges.

Workbook Exercise 5
Tell students that if they have trouble with this page, they can copy the figures and then fold to see if they can form the solid.

Activity 2.2b   **Nets of cubes**

1. Explore nets of cubes.
   - Refer to **workbook exercise 5, problem 1**, which students should have already completed for homework.
   - Ask students if they came up with any general ideas for determining if the figure was the net of a solid.
   - Divide students into groups. Give each group some square graph paper (you can copy p. 21 of this guide). Ask the students in each group to see if they can draw all the possible nets that form a cube. There are 11 possibilities, apart from rotations and reflections. This is hard, so guide students if they are having trouble finding the 11 different nets. They can cut out their figures and fold them into cubes.

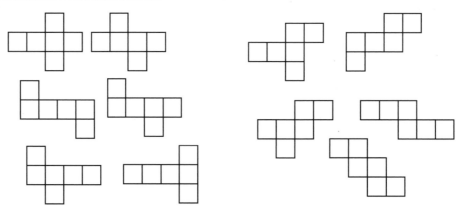

Copyright © 2006 SingaporeMath.com Inc., Oregon

- Guide students in listing some general rules for determining if a net of 6 squares could form a cube. They can cut out their 11 nets and try to categorize them into groups.
  - ➢ The longest panel cannot have more than 4 squares.
  - ➢ If the longest panel has 4 squares, the remaining 2 have to be placed on opposite sides.
  - ➢ If the longest panel has 3 squares, the remaining 3 can be placed on either side but there must always be at least 2 free edges on either side of the longest panel.
  - ➢ If the longest panel has 2 squares, the remaining 4 must be evenly distributed on either side such that there will be one free edge on either side of the panel.

Activity 2.2c   **Nets of other solids**

1. Explore nets of cuboids.
   - **Refer to workbook exercise 5, problem 2.**
   - Ask students to discuss any methods they used to determine if the nets could form a cuboid. One of the first things to look for, other than whether the net has 6 faces, is to determine if adjacent sides are equal in length. If they are not, the figure can't be the net of a cuboid.

2. Explore nets of other solids.
   - Have students do **task 2, textbook p. 19**. If necessary, they can trace the nets and see if they can fold them into solids. Figures A and D are nets of the solid.
   - Discuss with them why they could eliminate B and C. B does not have enough faces, and the bottom two edges of C are obviously of different lengths, so they won't be able to join along their lengths to form an edge.
   - You can have students do **workbook exercise 6** in class and discuss reasons for eliminating certain of the nets. Nets that do not have the correct number of faces of the correct shape, and ones where adjacent sides are not the same length can be eliminated.
   - If students have trouble with this exercise, have them trace and cut out the shapes to see if they fold up into the solid. That will help them see why adjacent sides need to be the same length.

3. Optional: Nets of tetrahedrons.
   - Tell students that a tetrahedron is a pyramid with four equilateral triangles. Show them a solid model of a tetrahedron.
   - Ask students to find all the possible nets for a triangular pyramid or tetrahedron. There are only two possibilities.

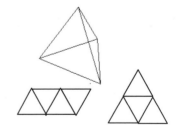

Workbook Exercise 6

Copyright © 2006 SingaporeMath.com Inc., Oregon

Activity 2.2d   **Solids from nets**

1. Find the solid of a net.
   - Discuss **task 3, textbook p. 20**. B is the solid formed from this net.
   - Have students explain why the others can't be formed from the net. The net has 6 faces, whereas A has 6 and C has 4. D has 4 triangles and 1 rectangle, whereas the net has 3 rectangles and 2 triangles.
   - Note that drawing 2-dimensional figures can sometimes make the angles that "go back" into the page look different from angles drawn flat on the page. In figure B, it looks like all three angles are acute (less than 90°), while the net has one slightly obtuse angle (greater than 90°). Students can trace the figure and fold it to see that when they put the largest square face towards them and hold it just below eye level, the angle farthest back does look smaller.

2. You can have students do **workbook exercise 7** in class and discuss reasons for eliminating certain of the solids.

3. Optional. You may want to give students some problems involving nets of cubes where they need to visualize where each face ends up on the resulting cubes. They should find the answer first by examining the net. They can then verify their answer, if necessary, by cutting out the net and folding. For example:

   (a) The figure here shows the net of a number cube. If the number cube is placed on the table with the "5" face down, what number is on the top of the cube? (1)

   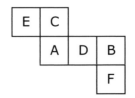

   (b) The figure here shows the net of a cube. If the cube is placed on the table with face "D" on top, which face is at the bottom of the cube? (E)

   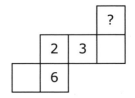

   (c) A cube has faces numbered 1, 2, 3, 4, 5, and 6. The numbers on the opposite faces of the cube add up to 7. The net of the cube is shown here. The numbers on 3 faces are not shown. Find the number on the face indicated by question mark. (1)

   (d) Two nets of the same cube are shown here. The second one has two blank faces. What letter goes in the face indicated by the question mark? (B)

   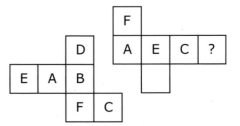

Copyright © 2006 SingaporeMath.com Inc., Oregon

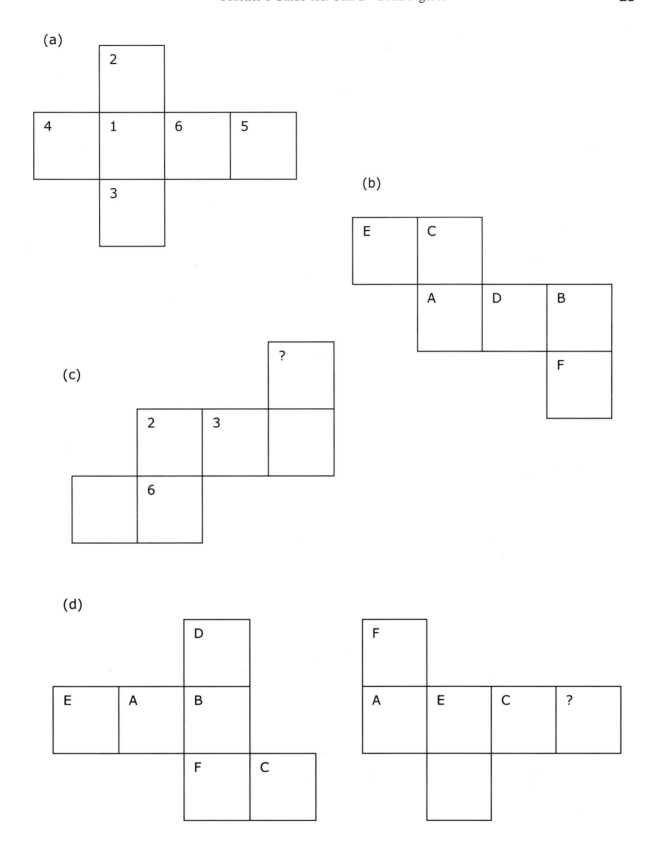

(a)

(b)

(c)

(d)

Unit 3 – Ratio

Objectives

- Compare quantities using ratios.
- Express a ratio in its simplest form.
- Relate ratios to units.
- Relate ratios to a fraction of a quantity.
- Relate proportion to ratios and fractions.
- Solve word problems involving ratios and proportion.
- Solve multi-step word problems involving changing ratios in before and after situations.

Suggested number of sessions: 12

	Objectives	Textbook	Workbook	Activities
Part 1 : Ratio and Fraction				**6 sessions**
12	▪ Compare quantities using ratios. ▪ Express a ratio in its simplest form. ▪ Relate ratios to units.	p. 21 pp. 22-23, tasks 1-5	Ex. 8	3.1a
13	▪ Review the use of pictorial models in solving word problems.			3.1b
14	▪ Express a ratio as a fraction of a quantity and vice versa.	pp. 24-25, tasks 6-8		3.1c
15		pp. 25-26, tasks 9-12 p. 29, Practice 3A: 1-3	Ex. 9	3.1d
16	▪ Solve word problems.	p. 27, tasks 13-14 p. 29, Practice 3A: 4-6	Ex. 10	3.1e
17		p. 28, tasks 15-16 p. 29, Practice 3A: 7-8	Ex. 11	3.1f
Part 2 : Ratio and Proportion				**2 sessions**
18	▪ Regard ratio as proportion. ▪ Solve word problems that involve direct proportions.	pp. 30-31 pp. 31-32, tasks 1-3	Ex. 12	3.2a
19	▪ Practice.	p. 33, Practice 3B		3.2b
Part 3 : Changing Ratios				**4 sessions**
20	▪ Solve word problems that involve finding a new ratio after a change.	p. 34 p. 35, tasks 1-2 p. 38, Practice 3C: 1, 7		3.3a
21	▪ Solve word problems where there is a transfer from one quantity to the other.	p. 36, tasks 3-4 p. 38, Practice 3C: 3, 5	Ex. 13	3.3b
22	▪ Solve ratio problems that involve only one changing quantity.	p. 37, task 5 p. 38, Practice 3C: 2,6,8		3.3c
23	▪ Solve ratio problems that involve more than one changing quantity.	p. 37, task 6 p. 38, Practice 3C: 4	Ex. 14	3.3d

Copyright © 2006 SingaporeMath.com Inc., Oregon

Part 1: Ratio and Fraction	6 sessions

Objectives

- Compare quantities using ratios.
- Express a ratio in its simplest form.
- Express a ratio as a fraction of a quantity.
- Express a fraction of a quantity as a ratio.
- Solve word problems which involve ratios, using pictorial models.

Materials

- *Primary Mathematics 5A* textbook and Teacher's Guide
- Connect-a-cubes or other cubes
- Counters

Homework

- Workbook Exercise 8
- Workbook Exercise 9
- Workbook Exercise10
- Workbook Exercise 11

Notes

In *Primary Mathematics 5A*, students learned to write ratios which involved two or three quantities. The also learned to find the simplest form of a ratio. If your students did not use *Primary Mathematics 5A*, you may want to do unit 5 of *Primary Mathematics 5A* along with what is covered here for pp. 21-23 of the *6A* textbook.

A ratio is a comparison of the relative size of two or more quantities. In a ratio, quantities can be compared without specifying the unit, as long as the same unit is used for both quantities. *In measurement, a unit is required.* A rope that is 6 *feet* long is not the same length as one that is 6 *meters* long. However, one rope can be accurately compared to another rope without specifying the unit. For example, one rope can be specified as twice as long as the other. The ratio of their lengths is then 2 : 1 whether the ropes are measured in feet, meters, inches or centimeters.

Equivalent ratios are ratios where the relative sizes of the quantities remain the same, but the unit is different. 200 : 300, 20 : 30, 10 : 15, and 2 : 3 are equivalent ratios. Equivalent ratios can be found by multiplying or dividing each term by the same number. If the terms of a ratio have a common factor, we can then simplify the ratio by dividing each term by the common factor. If there is no such common factor, the ratio is already in its *simplest form.* 2 : 3 is a ratio in its simplest form.

The importance of the concept of "unit" is constantly emphasized in *Primary Mathematics*. A unit can be a "one" or "one collection of several items". If the unit is 1 vegetable, the ratio of 200 carrots to 300 onions is 200 : 300. But if the unit is 100 vegetables, the ratio of 200 carrots to 300

Copyright © 2006 SingaporeMath.com Inc., Oregon

onions is 2 : 3. Notice that when we changed the unit to mean 100 vegetables, in effect we were finding the 2 : 3 as an equivalent ratio for 200 : 300. Usually, we use an equivalent ratio which is in its simplest form.

In *Primary Mathematics 5A*, students learned to solve whole-number word problems using part-whole and comparison models, and they worked with fraction and ratio problems using unit bar models. In this unit, they will be combining these concepts. If some of your students come from classes which did not use *Primary Mathematics 5A*, you will need to take time to cover this material from *Primary Mathematics 5A*.

When comparing quantities, we often use fractions. For example, one quantity can be $\frac{2}{3}$ as large as another. Since a ratio also compares quantities, a ratio can be converted into a fraction, or a fraction can be converted into a ratio.

For example, the ratio of the length of A to B is 2 : 3. If B is taken as the whole, we can say that the length of A is $\frac{2}{3}$ *of* the length of B. Note that the quantity after the "of" is the quantity that is considered the whole. The length of A is a fraction *of* the length of *B*. Note also that the total is the denominator (bottom) of the fraction.

A is $\frac{2}{3}$ of B

B is $\frac{3}{2}$ of A

A is $\frac{2}{5}$ of A + B

B is $\frac{3}{5}$ of A + B

If we take the length of A as the whole instead, then we can express the length of B as a fraction *of* the length of A. Or, we can take the sum of A and B as the whole (total) and express each length as a fraction of the total length.

In the problems in this section, the whole won't always be given last in the sentence. For example, the student may be asked to express the length of A as a fraction *of the length of B*, or be asked to find what fraction *of the length of B* is the length of A. In both of these cases, the whole is the length of B, and the answer is the same.

If we are given the fraction of one quantity relative to the other quantity, we can convert this fraction into ratio. If the length of A is $\frac{2}{3}$ of the length of B, we can diagram this relationship and easily see that the ratio of the length of A to the length of B is 2 : 3.

In word problems which involve ratio, we use the ratio along with other information to determine the unit and thus the quantity. We can draw bar models with units that represent the relationship between the quantities. For example:

Copyright © 2006 SingaporeMath.com Inc., Oregon

There are $\frac{5}{3}$ as many boys as girls. There are 20 more boys than girls.

We draw the relationship between the boys and girls as units. We see that there are 2 more units of boys than girls. Since there are 20 more boys than girls, one unit is 10.

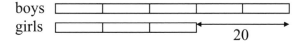

Once we find the value of one unit, we can find the answer to a variety of questions, such as how many children there are, or how many boys there are, or how many girls there are.

When discussing the problems in the text, you can present each problem to the students and guide them in drawing diagrams without having them first look at the diagram in the text ahead of time. They might not draw their diagram in the same way it is shown in the text, and this can be a point for discussion and learning. They may approach the problem in a different, but valid, way. Allow students the opportunity to explain their solutions (orally). Solutions in the text (or in this guide) are not the only correct approach. It is important for students to appreciate that there is not necessarily a formulaic approach for solving each word problem.

Keep in mind that diagramming is a problem solving tool, and not the *only* problem solving tool. If the solution is obvious to a student without a diagram, don't insist that he or she draw one. However, make sure students understand the tool and how to use it, and encourage solutions that involve diagramming in class discussions.

Copyright © 2006 SingaporeMath.com Inc., Oregon

Activity 3.1a   **Ratio**

1. Review ratios. Use counters or drawings.
   - Show 6 counters of one color (such as blue) and 12 of another (such as red). Remind students that when we compare quantities, we can use ratios. There are 6 blue counters for 12 red counters, so the ratio of blue to red counters is 6 : 12.
   - Now ask students for the ratio of red to blue counters. Lead them to see that it is 12 : 6. The correct order must be used in giving the ratio of one quantity to another.
   - Group the counters by 2, 3, and 6 to show that the unit can be 1 counter, 2 counters, 3 counters, or 6 counters. You can draw unit bars under the counters to show how unit bars relate to ratios. Each unit must the same value to relate the two quantities.

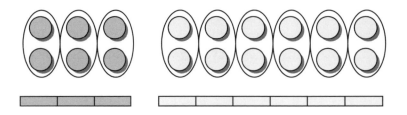

1 unit = 2 counters
Ratio of blue counters to red counters = 3 : 6

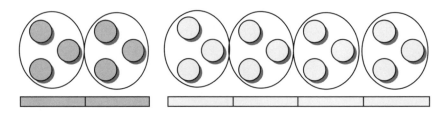

1 unit = 3 counters
Ratio of blue counters to red counters = 2 : 4

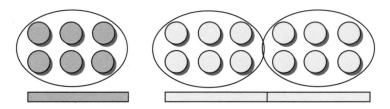

1 unit = 6 counters
Ratio of blue counters to red counters = 1 : 2

   - Remind students that 6 : 12, 3 : 6, 2 : 4, and 1 : 2 are equivalent ratios. They all relate the number of blue counters to the number of red counters. Lead students to see that the different equivalent ratios were derived from the first ratio by dividing both terms by the same number.

$$6 : 12 = 3 : 6 = 2 : 4 = 1 : 2$$

Copyright © 2006 SingaporeMath.com Inc., Oregon

- Remind students that the ratio 1 : 2 is the *simplest form* of that ratio. For every 1 blue counter, there are 2 red counters.
- Ask students to simplify 60 : 40 : 200. Point out that simplification can be done in several steps. We can divide each term by an easily determined common factor (such as 2 until one number is no longer even, then 3 or 5). Or, they can simplify in one step using the greatest common factor.

$$60 : 40 : 200 = 30 : 20 : 100 \text{ (common factor 2)}$$
$$= 15 : 10 : 50 \text{ (common factor 2)}$$
$$= 3 : 2 : 10 \text{ (common factor 5)}$$

$$60 : 40 : 200 = 3 : 2 : 10 \text{ (common factor 20)}$$

- For additional review, refer to **unit 5, part 1, pp. 71-76 and part 3, pp. 80-81 in *Primary Mathematics 5A*.**

2. Discuss the contents of **p. 21 in the textbook**.
   - Each unit is $10. Lead students to see that 3 : 2 means we are comparing 3 units to 2 units. We could also express the ratio as 30 : 20, or 15 : 10. However, 3 : 2 is the simplest form. For every $3 Susan spent, Mary spent $2.

3. Discuss **tasks 1-2, textbook p. 22**.

4. Discuss **tasks 3-5, textbook p. 23**. Students should express the ratios in their simplest form.

5. For additional review, refer to ***Primary Mathematics 5A*, p. 71, pp. 72-73, tasks 1-6, p. 75, p. 76, tasks 1-2, p. 80, and p. 81, task 1.**

   Workbook Exercise 8

Activity 3.1b   Word problems

1. Review using bar diagrams with units to solve word problems. Refer to the *Primary Mathematics 5A* textbook, unit 1, part 7; unit 3, part 7; and pages 77-79 and 81-82, along with the corresponding material in the *Primary Mathematics 5A Teacher's Guide*.

Copyright © 2006 SingaporeMath.com Inc., Oregon

Activity 3.1c   **Ratio and fractions**

1. Illustrate the relationship between ratios and fractions.
   - Use cubes, blocks, or draw pictures on the board. Display 2 blocks of one color, such as yellow, and 3 of another, such as red.
     - o Ask students to express the number of yellow blocks as a fraction of the total number of blocks. The number of yellow blocks is $\frac{2}{5}$ of the total number of blocks.
     - o Add two more red blocks, and move the yellow ones to above the (now 5) red blocks. Ask students to express the yellow blocks as a fraction of the total number of blocks. The number of yellow blocks is $\frac{2}{7}$ of the total number of blocks.
     - o Now ask students to express the number of yellow blocks as a fraction of the number of red blocks. The number of yellow blocks is $\frac{2}{5}$ of the number of red blocks.
     - o Point out that the denominator (bottom) of the fraction now refers to the number of red blocks (5) instead of all the blocks (which is now 7). We are finding yellow blocks as a fraction of red blocks, not as a fraction of total blocks.
     - o Ask students for the ratio of yellow blocks to red blocks (2 : 5). Since the number of yellow blocks is $\frac{2}{5}$ the number of red blocks, we can say that the ratio of yellow blocks to red blocks is 2 : 5.
   - Display 2 yellow and 2 red blocks.
     - o Ask students to compare the red blocks to the yellow blocks. There are as many red as yellow blocks, so the ratio of the number of yellow blocks to the number of red blocks is 1 : 1.
     - o Now add a red block and ask students to compare the *red* blocks to the *yellow* blocks. There are one and a half times as many red as yellow blocks. We can say that the number of red blocks is $\frac{3}{2}$, or $1\frac{1}{2}$, of the number of yellow blocks. There are $\frac{3}{2}$ as many red blocks as yellow blocks, and the ratio of the number of red blocks to the number of yellow blocks is 3 : 2.
     - o Add 2 more red blocks. Ask students student to express the number of *red* blocks as a fraction of the number *of yellow* blocks. The number of red blocks is $\frac{5}{2}$ of the number of yellow blocks.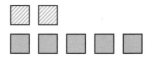
     - o Now ask for the ratio of red to yellow blocks (5 : 2). If we are told that the ratio of red blocks to yellow blocks is 5 : 2, then we can also say that the number of red blocks is $\frac{5}{2}$ *of the number of yellow blocks.*
     - o The number of yellow blocks is $\frac{2}{5}$ *of the number of red blocks.*

Copyright © 2006 SingaporeMath.com Inc., Oregon

- Tell students that each block can be diagrammed as a unit. Draw a bar diagram.
- Tell students that we are told that the ratio of yellow flowers to red flowers in a bouquet is 2 : 5. Ask them the following:
  - The number of yellow flowers is what fraction of red flowers? The number of yellow flowers is $\frac{2}{5}$ *of the number of red flowers*. Note that since we are comparing yellow flowers *to* red flowers, the number of red flowers is in the denominator of the fraction.
  - Express the number of yellow flowers as a fraction of total flowers in the bouquet. Yellow flowers are $\frac{2}{7}$ of all the flowers in the bouquet. This time the total we use is the number of units for all the flowers (7), so that is what is in the denominator (bottom).

2. Discuss **task 6, textbook p. 24**.
   - Go over each statement and illustrate line by line. Emphasize which quantity is to be taken as the "whole" and so appears in the denominator of the fraction.

3. Have students do **tasks 7-8, textbook p. 25** and explain their solutions.
   - In task 7, you can also ask students to express the number of boys as a fraction of the total number of children.
   - For task 8, also ask for Jim's money as a fraction of Samy's money.

Activity 3.1d   **Ratio and fractions**

1. Discuss **tasks 9-12, textbook pp. 25-26**.
   - For task 9, point out that in the questions asked for (c) and (d), the quantity to be taken as the whole is given first in the sentence. Tell students that (c) could also be phrased as: "Meihua's savings is what fraction of Sumin's savings?" or "Express Meihua's savings as a fraction of Sumin's savings." In either case, Sumin's savings is considered the whole, and is the value of the denominator.
   - Ask: What fraction of their total savings is Meihua's savings?
     - Method 1: $\frac{420}{420+350} = \frac{420}{770} = \frac{6}{11}$ Meihua's savings is $\frac{6}{11}$ of the total of both their savings.
     - Method 2: If we have already found the ratio of Meihua's savings to Sumin's savings to be 6 : 5, we can compute the total units, 6 + 5 = 11. The ratio of Meihua's savings to their total savings is 6 : 11, so the fraction of Meihua's savings is $\frac{6}{11}$ of their total savings.

2. Have students do some problems where they need to draw the models themselves.
   - You can use **problems 1-3, Practice 3A, textbook p. 29**.
   - Have students share their diagrams and solutions. Discuss any alternate solutions students may have found.

Workbook Exercise 9

Copyright © 2006 SingaporeMath.com Inc., Oregon

Activity 3.1e   **Word Problems**

1. Discuss **tasks 13-14, textbook p. 27**.
   - First relate the information in the diagram to the information given in the word problem.
   - Tell students that in word problems which involve ratio, we are given a value that we can assign to one or more units. Then we solve for 1 unit, which lets us find the value for any number of units.
   - Have students solve first for 1 unit, then for 15 units.

2 units = \$60
1 unit = \$$\frac{60}{2}$ = \$30
15 units = \$30 x 15 = \$450

2. Have students do some problems where they need to draw the models themselves.
   - You can use **problems 4-6, Practice 3A, textbook p. 29**.
   - Have students share their diagrams and solutions. Discuss any alternate solutions students may have found.

Workbook Exercise 10

Activity 3.1f   **Word Problems**

1. Discuss **task 15, textbook p. 28**.
   - As the textbook shows, for this problem we first draw the bars for Jim and Raju and show that the ratio for Jim's marbles to Raju's marbles is 2 : 1. Then, we draw bars to show the ratio between Raju's and Lihua's marbles, but we also show that since the number of Raju's marbles does not change, the bar is drawn the same length in both cases. From this, we can see that if we divide the longer units in the first two bars into fourths, then all the bars have equal units and we can find the ratio between the numbers of marbles for all three boys. Essentially, we are finding an equivalent ratio for Jim's marbles to Raju's marbles where the number of units for Raju is the same as in the number of units in the ratio for Raju's marbles to Lihua's.

     Jim : Raju   = 2 : 1
       Raju : Lihua =   4 : 5
                    ↓
     Jim : Raju   = 8 : 4
       Raju : Lihua =   4 : 5
                    ↓
     Jim : Raju : Lihua = 8 : 4 : 5

   - Discuss the problem shown at the right where the equivalent ratios for *both* quantities need to be found. Have students draw diagrams

     In this case, the number of Peter's marbles is the same in both ratios. To compare all three quantities, we find equivalent ratios in which the number of units for Peter is the same.

     > The ratio of Peter's marbles to Paul's marbles is 2 : 3 and the ratio of Peter's marbles to Mary's marbles is 5 : 2. What is the ratio of Peter's marbles to Paul's marbles to Mary's marbles?

Peter : Paul = **2** : 3 = **10** : 15

Peter : Mary = **5** : 2 = **10** : 4

Peter : Paul : Mary = 10 : 15 : 4

The ratio of Peter's marbles to Paul's marbles to Mary's marbles is 10 : 15 : 4.

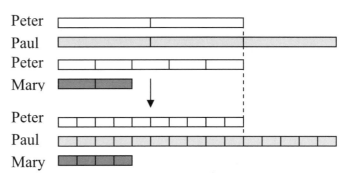

2. Discuss **task 16, textbook p. 28**.
    *   Then, discuss the word problem shown at the right. Have students draw a diagram.
        Here, the total of Diego's money is one whole, and the total of Alfonso's money is another whole. We need to find the fraction of Alfonso's money to Diego's money, which will be the new whole. To solve this problem, we will need to find a common multiple of the two numerators.

        Alfonso's money is $\frac{9}{10}$ of Diego's money.

$\frac{3}{5}$ of Diego's money is the same amount as $\frac{2}{3}$ of Alfonso's money.

What fraction of Diego's money is the same as all of Alfonso's money?

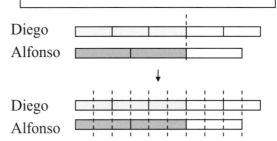

3. Have students do **problems 7-8, Practice 3A, textbook p. 29**. Possible solutions are shown here.

#7.

men = 5 units
women = 8 units

There are 3 more units of women than men. Total units = 5 + 8 = 13

   3 units = 24

   1 unit = $\frac{24}{3}$ 24 ÷ 3 = 8

   13 units = 8 × 13 = 104

There are 104 workers.

#8. (a)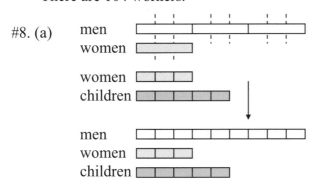

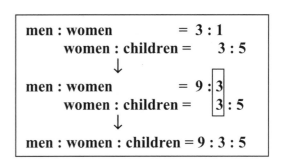

men : women   = 3 : 1
   women : children =   3 : 5
                    ↓
men : women   = 9 : 3
   women : children =   3 : 5
                    ↓
men : women : children = 9 : 3 : 5

Copyright © 2006 SingaporeMath.com Inc., Oregon

(b)   children = 5 units; total people = 17 units
      5 units = 20
      1 unit $= \dfrac{20}{5} = 4$
      17 units $= 4 \times 17 = 68$
      There are 68 people.

4. Optional: Write the problem shown here on the board. Guide students in drawing a diagram and solving the problem.

> $\dfrac{3}{4}$ of Sam's weight is the same weight as $\dfrac{4}{5}$ of Tom's weight. What is the ratio of Sam's weight to Tom's weight?

*   We can first draw 4 units for Sam's weight. Then we can draw a vertical line to mark off 3 of them.

*   These 3 units are equal to $\dfrac{4}{5}$ of Tom's weight. So, for Tom's bar, we draw 4 units to line up with the 3 in Sam's bar, and add one more to make 5 sections in Tom's bar. We need to compare 3 of Sam's units with 4 of Tom's unit. Student's drawings need to be correct, but they don't need to be as neat as the one shown here.

*   Now, in order to find a ratio, we need equal-sized units. Ask for suggestions. We have 3 of Sam's units equaling 4 of Tom's units. We can divide each of Sam's units into fourths and each of Tom's units into thirds. This gives 12 units for the three of Sam's and the four of Tom's. Note that 12 is a common multiple of 3 and 4.

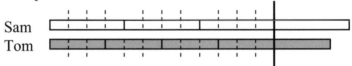

*   We then need to divide the remaining portions of both their bars.

*   The ratio of Sam's weight to Tom's weight is 16 : 15.

Workbook Exercise 11

Copyright © 2006 SingaporeMath.com Inc., Oregon

Part 2: Ratio and Proportion   **2 sessions**

Objectives

- Express proportion as a fraction or a ratio.
- Solve word problems involving direct proportion.

Materials

- Recipes, maps

Homework

- Workbook Exercise 12

Notes

When two ratios are equivalent, we say that they express the same *proportion*. As, 3 : 5 = 15 : 25 states that 3 : 5 has the same proportion as 15 : 25. Since we can express ratios as fractions, the equivalence of two ratios can also be expressed in the form of fractions. 3 : 5 can be written as $\frac{3}{5}$, and 15 : 25 can be written as $\frac{15}{25}$. We already know that $\frac{3}{5} = \frac{15}{25}$, and that we call these equivalent fractions. In the same way, both the ratio 3 : 5 and the ratio 15 : 25, being equivalent, express the same proportion.

Proportions are useful with such things as maps, cooking, and scale drawings. If the scale of a map is 1 cm per 100 km, then a distance of 20 cm on the map would represent an actual distance of 2000 kilometers. To make 5 cakes, we multiply each ingredient of the cake recipe 5 times. When the recipe calls for 3 cups milk and 5 cups of flour, then we will use 15 cups of milk and 25 cups of flour, so that those ingredients would still be in the same proportion.

Problems involving proportion can be represented with unit drawings, in the same way as in problems involving ratios. If we know the ratio, and the value of one quantity, the value of the total, or the difference between the quantities, we can then find the value of one unit. From that, we can find the value of the second quantity.

For example, mix sand and cement in a ratio of 3 : 5 to make concrete, as with 3 buckets of sand and 5 buckets of cement. How many buckets of cement would we mix with 15 buckets of sand to keep the same proportion? We can draw a bar diagram of the ratio, and see that 15 buckets of sand must equal 3 units. So one unit equals 5 buckets, and then 5 units of cement equal 25 buckets. In other words, we need to mix 25 buckets of cement with the 15 buckets of sand to make cement using the same sand-cement proportion.

3 units = 15 buckets
1 unit = 15 ÷ 3 = 5 buckets
5 units = 5 × 5 = 25 buckets

We can also find the answer using equivalent fractions: $\frac{3}{5} = \frac{15}{?}$

Activity 3.2a   **Proportion**

1. Discuss **textbook pp. 30-31**.
   - Point out that when we increase the number of buckets of cement, the number of buckets of sand should also be increased by the same factor.
   - The amounts of cement and of sand are kept in the same *proportion*.

     $\dfrac{\text{Number of buckets of cement}}{\text{Number of buckets of sand}}$ is always equivalent to $\dfrac{5}{3}$ when expressed in its simplest form.

   - Ask students to solve these proportions by finding equivalent fractions.
   - Lead students to see that a proportion can be described by using ratios (top of p. 31).
   - Discuss other examples which involve proportions. For example:
   - ➤ $\dfrac{\text{weight on the moon}}{\text{weight on earth}} = \dfrac{1}{6}$

     Have students find what their weight would be if they were standing on the moon.
   - ➤ Discuss the scale on a map, and have students estimate some actual distances between cities or other features. Have them use the map and a ruler. Emphasize that they are applying the concept of proportions.

2. Discuss **tasks 1-3, textbook pp. 31-32**.
   - Students should see that we can solve a problem that involves proportion as a ratio problem.
   - In task 2(b), we are given the value of 1 unit, and can find the value of 3 units using multiplication.
   - In task 2(c), we are given the value of 3 units, and can find the value of 1 unit using division.
   - In task 3, have students first show the value for 1 unit, and then find the answer to the problem.

       3. (a)   3 units = 12 ℓ   (b)   5 units = 10 ℓ

                    1 unit = $\dfrac{12}{3}$ ℓ = 4 ℓ   1 unit = $\dfrac{10}{5}$ ℓ = 2 ℓ

                    2 units = 4 ℓ × 2 = 8 ℓ   3 units = 2 ℓ × 3 = 6 ℓ

Workbook Exercise 12

Activity 3.2b   **Practice**

1. Have students work on the problems in **Practice 3B, textbook p. 33**, and share their solutions. All these problems can be solved by first finding the value for 1 unit, as shown here for problem 9:

        total units = 4 units + 5 units + 6 units = 15 units

        15 units = 60 cm

        1 unit = $\dfrac{60}{15}$ = 4 cm

        4 units = 4 cm × 4 = 16 cm

        The shortest side is 16 cm.

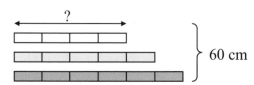

Copyright © 2006 SingaporeMath.com Inc., Oregon

Part 3: Changing Ratios   4 sessions

Objectives

- Solve word problems which involve ratios or proportions that change according to a "before" and "after" concept.

Homework

- Workbook Exercise 13
- Workbook Exercise 14

Notes

A changing-ratio problem involves a "before" situation and an "after" situation which brings about a change in the ratio of the two quantities. These problems can be illustrated by drawing separate diagrams for each situation. Generally with these problems, as with the ratio and proportion problems already seen in this unit, solutions begin with finding the value of one unit in either in the "before" situation or in the "after" situation.

Determining the unit is the key to solving these problems. The diagram must always indicate the quantity that does not change, if there is one, since this will also be a key factor in solving the problem. Or, if both quantities change, the amount by which they change should be shown.

Changing ratio problems are often challenging for students, and generally require diagrams in order to solve them (without resorting to the algebraic equations they will learn in higher grades). The examples in the textbook show a final diagram. Don't just have students look at these diagrams. Through discussion, lead the students in guiding you in drawing them on the board so that they can understand the thinking process involved in setting up the diagram.

Copyright © 2006 SingaporeMath.com Inc., Oregon

Activity 3.3a   **Finding a new ratio**

1. Discuss the problem on **textbook p. 34**, going over it in detail.
    -   In this problem, one quantity changes, but the other does not. The unchanging quantity provides the key for relating units in the "before" situation to units in the "after" situation.
    -   Guide students, or let them guide you, in drawing the diagram step-by-step, discussing the thought process.
        o   The problem states that the ratio of Peter's stamps to Henry's stamps is 2 : 3. We can draw the diagram to show that ratio. Draw two equal units for Peter's bar, and 3 for Henry's bar.
        o   Then, the number of Peter's stamps is increased by 8, while the number of Henry's stamps stays the same. So for our "after" diagram, we draw a bar for Henry's stamps, the same length as Henry's bar in the "before" situation. Then we draw Peter's bar, which has now become a little longer. (It helps to line up all the bars, drawing a vertical line from the edges of the bars in the "before" diagram to the bars in the "after" diagram.)
        o   The problem states that the ratio of Peter's stamps to Henry's has become 5 : 6. Divide each of the 3 units of Henry's "after" bar in half. (That is, Henry's bar goes from a bar of 3 units to one the same length but now of 6 units.)
        o   Now divide up Peter's bar. Since each of the "before" units is now halved, that part of his bar which is the same length as in the "before" situation is now 4 units. We can see that the fifth unit (since the ratio is 5 : 6) must stand for the 8 stamps that Peter bought. We now have a value for one (after) unit: 8 stamps.
        o   From the value for 1 (after) unit, we can determine how many stamps Peter had at first. 8 stamps by 4 (after) units = 32 stamps.
    -   Sometimes, the problem is made clearer when the "after" diagram is drawn "on top" of the "before" diagram, particularly after students have become familiar with these types of problems. We could have simply drawn the "before" diagram, then split Henry's units into half to get 6 units (since the new ratio has 6 units for Henry), done the same with Peter's units, and added a unit to show the new ratio of 5 : 6. Since the increase must be the 1 unit we added, the new units have a value of 8. If students still have trouble seeing the relationship between the "before" and "after" diagram, try superimposing them.

2. Discuss **tasks 1-2, textbook p. 35**.
    -   In these problems, a change occurs in the amount and the student needs to find a new ratio. We are given the original ratio and a value that allows us to find the actual number each person has. Guide the students in finding the amount each person has, changing the amount according to the information in the problem, and then finding the new ratio.
    -
        #1. (In the 3rd edition, Joe is Ali, and Damon is Gopal.)
             We use the original ratio (2 : 7), and the value we are given (Damon has 56 stamps) to find the number of stamps Joe has. Discuss each "before" and "after" step needed to find the new ratio. Draw the book's bars on the board as you go through the problem. In this problem, we are using new actual values to find a new ratio, so it is not necessary to draw an "after" diagram.

Copyright © 2006 SingaporeMath.com Inc., Oregon

#2.   (In the 3rd edition, Matthew is Minghua and Susan is Sulin)
Draw bars on the board as you go through this problem.

Before:
Matthew has 2 units, and Susan has 3 units. Matthew has 40 books.
2 units = 40

1 unit = $\frac{40}{2}$ = 20

3 units = 20 × 3 = 60 (Number of Susan's books)

After:
Number of Mathew's books = 48
Number of Susan's books = 60 (the number doesn't change)
New ratio = 48 : 60 = 4 : 5

3.   Have students do **problems 1 and 7 of Practice 3C, textbook p. 38** and share their solutions.
Possible solutions:

#1.

Before:
1 unit = 6
number of boys = 4 units = 6 × 4 = 24
number of girls = 3 units = 6 × 3 = 18

After 2 girls join the choir:
Number of boys = 24 (unchanged)
Number of girls = 18 + 2 = 20
New ratio of boys : girls = 24 : 20 = 6 : 5

#7.
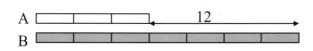

(a)   Ratio = 3 : 7
4 units = 12 marbles

1 unit = $\frac{12}{4}$ = 3 marbles

10 units = 10 × 3 = 30
There are 30 marbles altogether.

(b)   *Before:*
1 unit = 3 marbles (as per the answer in (a))
Number of marbles in A = 3 × 3 = 9
Number of marbles in B = 3 × 7 = 21

After:
Moved 3 marbles from A to B:
Number of marbles in A = 9 − 3 = 6
Number of marbles in B = 21 + 3 = 24
New ratio = 6 : 24 = 1 : 4

Copyright © 2006 SingaporeMath.com Inc., Oregon

Activity 3.3b   **Transferring from one quantity to another**

1. Discuss **task 3, textbook p. 36**.
    *   In this problem, we need to find a new ratio, but we are not given any values. However, we are given an initial ratio and told how the total amounts change.
    *   Draw the first diagram on the board (the "before" diagram).
    *   Ask students what changes. The problem states that half the marbles are moved from A to B. Since A has 4 units, then 2 units are moved from A to B. Show this when drawing the second diagram on the board (the "after" diagram).
    *   Note that a key to solving this problem is that the total number of marbles and the total number of units do not change – there is just a transfer from one bar to the other.
    *   You may want to discuss an alternate diagram. Since the total does not change, we can show this by putting the bars next to each other, rather than aligning them vertically.

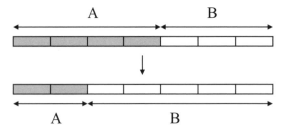

2. Discuss **task 4, textbook p. 36**.
    *   Point out that in this problem, not only does the total amount not change (an amount is transferred from one quantity to the other), but after the transfer, the quantities are equal.
    *   Lead students to see that in order for the quantities to become equal, we transferred half the difference from the larger quantity to the smaller quantity.
    *   Since half the difference is 1 unit, and Susan gave $20 to Mary, then 1 unit must be $20.
    *   The diagram in the text is the final diagram. When drawing the problem on the board, step-by-step, we do not know that 1 unit is $20 until the last step of the drawing.

3. Have students do **problems 3 and 5 of Practice 3C, textbook p. 38** and share their solutions. Possible solutions:

    #3. In this problem we start out with equal quantities (ratio is 1 : 1). Then, after a transfer from one to the other, we are given the new ratio (7 : 5). It is generally helpful to diagram the situation in which we are given a ratio other than 1 : 1 first (in this case the "after" situation). The "after" ratio is 7 : 5, which is a total of 12 units. Since they started ("before") with an equal number of units, each had half of the 12 units, or 6 units each. Since Samy went from 6 units to 5, he gave 1 unit to Ali. That tells us that 1 unit is $15.

1 unit = $15
Both had 6 units at first.
6 units = $15 × 6 = $90
Both had $90 at first.

Copyright © 2006 SingaporeMath.com Inc., Oregon

#5. We are told that the number of Jim's stamps is $\frac{3}{4}$ of the number of David's stamps. We can convert this to a ratio (3 : 4) and diagram the "before" situation. That gives Jim 3 units and David 4 units. Since Jim gives David half of his stamps, he gives David $1\frac{1}{2}$ units. Jim ends up with $1\frac{1}{2}$ units and David ends up with $5\frac{1}{2}$ units. Since ratios should be expressed in whole numbers, we can divide each unit in half to make 2 units for each, so that the ratio is now 3 : 11.

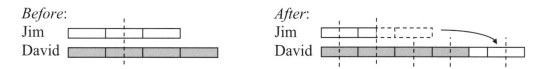

4. Optional: Give the following challenging problem to students as homework or to work on in groups:

Abe, Ben, and Casey shared some trading cards in the ratio 3 : 6 : 4. Abe kept $\frac{1}{5}$ of his cards and gave the rest of his cards to Ben and Casey according to the ratio 1 : 3. As a result, Ben had 60 cards more than Casey.
(a) How many cards did Abe keep for himself? (Ans: 45)
(b) Find the total number of cards Ben had after receiving cards from Abe. (Ans: 495)

If students need a hint, tell them that they need to find an equivalent ratio for 3 : 6 : 4 so that Abe can keep $\frac{1}{5}$ of his cards (15 : 30 : 20). Then the problem becomes simply transferring units, finding the difference in number of units between Ben and Casey (4) and setting that equal to 60.

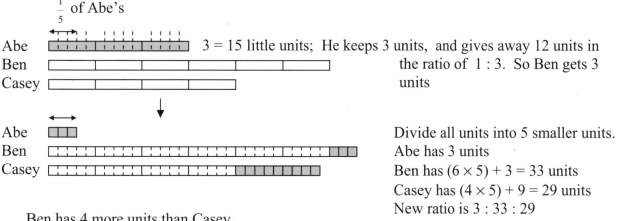

Ben has 4 more units than Casey.
4 units = 60 cards
1 unit = 60 ÷ 4 = 15 cards

(a) Number of cards Abe keeps = 3 units = 3 × 15 = 45 cards
(b) Number of cards Ben has = 33 units = 33 × 15 = 495 cards

Copyright © 2006 SingaporeMath.com Inc., Oregon

Workbook Exercise 13

In problem 1 of workbook exercise 13, the student can find the "before" amount, then the "after" amount, and then the new ratio. In problem 2 no amounts are given, but since we are given the fraction of the first amount that changes, we can find the new ratio.

For the first problem, you may want to discuss an alternate solution. How can we change the units for Suhua's stickers so that she gives Meili one fourth of the units? If we use the equivalent ratio 12 : 28, Suhua can give Meili one fourth of the units, or 7 of the 28. So Meili's units become 12 + 7 = 19, and Suhua's units become 28 – 7 = 21, and the new ratio is 19 : 21. In this solution, we don't need the information that Suhua has 32 more stickers than Meili.

Activity 3.3c   **Ratio problems where one quantity changes**

1. Discuss **task 5, textbook p. 37**. (In the 3rd edition, Ian is Ali and Juan is Gopal)
   - In this problem, Ian's money does not change.
   - A shortcut way of diagramming this problem is to draw the ratio for the "before" situation only, and then crossing off $\frac{1}{2}$ of Juan's bar, dividing all the units in half, and labeling the difference ($60). For some students, this can be easier to visualize, or to draw, since Juan gets rid of half of his money, which is half of his bar. This leaves him with $1\frac{1}{2}$ units, which we then turn into 3 half-sized units. Students who have trouble drawing neatly and lining up units might prefer this method. You may want to show this on the board. The final diagram could be something like this:

1 after-unit = $60
4 after-units = $60 × 4 = $240
Ian has $240

2. Have students do **problems 2, 6 and 8 of Practice 3C, textbook p. 38** and share their solutions. Possible solutions:

#2.   *Before*:

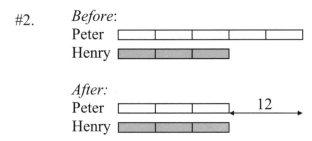

2 unit = $12
1 unit = $\frac{12}{2}$ = $6
Peter had 5 units at first.
5 units = $6 x 5 = $30
Peter had $30 at first.

Copyright © 2006 SingaporeMath.com Inc., Oregon

#6.   *Before*:

Meili ▭

Sulin ▭▭

After:

(a)   Meili starts with 1 unit, and Sulin starts with 2 units (1 : 2 ratio). Then Meili adds 12 books, and the ratio becomes 2 :1, but Sulin's books stay the same. In the "after" diagram Sulin's bar is the same (2 units), but Meili now has twice as many units as Sulin (4 units). This new ratio (4 : 2) is the same as 2 : 1..

(a)   3 units = 12 books

1 unit $= \dfrac{12}{3} = 4$ books

Sulin had 2 units.

2 units $= 4 \times 2 = 8$ books

Sulin had 8 books.

(b)   Meili now has 4 units

4 units $= 4 \times 4 = 16$ books

Then, Sulin adds 5 books.

Sulin then has $8 + 5 = 13$ books

New ratio = 16 : 13

#8.   John

Sumin

John begins with 5 units. Then he spends half of his money, so ends up with $2\frac{1}{2}$ units. We can divide the each unit into half to get the equivalent ratio 10 : 4. John spends 5 of these half-units and has 5 half-units left. Since Sumin still has 2 units (4 half-units), John has 1 more half-unit than Sumin. He has $20 more than Sumin, so 1 half-unit = $20.

(a)   Amount of money Sumin has = 4 units $= \$20 \times 4 = \$80$

(b)   Amount of money John had at first = 10 units $= \$20 \times 10 = \$200$

3.   Optional: Give the following challenging problem to students as homework or to work on in groups:

The ratio of the number of nails to the number of screws in a tool box is 11 : 9. Tom uses 70 nails and then adds in 300 nails. The final ratio of the number of nails to the number of screws is 5 : 2. How many screws are there? (Ans: 180)

If students need a hint, point out that the number of screws doesn't change. So we can start by finding equivalent ratios for 11 : 9 and 5 : 2 which have the same number for screws (22 : 18 and 45 : 18.)

Copyright © 2006 SingaporeMath.com Inc., Oregon

Activity 3.3d   **Ratio problems where more than one quantity changes**

1. Discuss **task 6, textbook p. 37**.
   - In this problem, both quantities change. The key to this problem is that both quantities are equal to begin with.
   - In order to help students visualize that 1 unit is the difference between $25 and $18 in this problem, you may want to diagram the "Before" and "After" situation on the board a little differently, superimposing the two situations. That is, draw the two equal bars, and then show 2 units on John's, with $25 making up the rest of his bar, and 3 units on the Matthew's, with $18 making up the rest of his bar. (In the 3rd edition, Matthew is Minghua.)

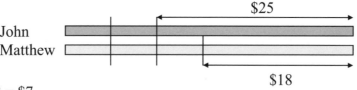

   1 unit = $25 – $18 = $7
   John has 2 units + $25, or $7 × 2 + $25 = $14 + $25 = $39
   (or, Matthew has 3 units + $18 = $7 × 3 + $18 = $21 + $18 = $39)
   They both had $39 at first.

2. Have students do **problem 4 of Practice 3C, textbook p. 38** and share their solutions. Problem 4 is similar to task 6.

3. Provide additional practice with a variety of ratio problems:
   - Kate's money is $\frac{5}{7}$ of Bianca's money. If Bianca gives $\frac{1}{2}$ of her money to Kate, what will be the ratio of Kate's money to Bianca's money? (17 : 7)

   - The ratio of Sharon's savings to Alice's savings is 3 : 8. Sharon spent $\frac{1}{3}$ of her savings and Alice spent $70 of her savings. The final ratio of Sharon's savings to Alice's savings is 4 : 9. How much did Sharon have at first? ($60)

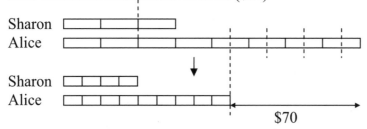

   After:
   7 units = $70
   1 unit = $10
   Sharon had 6 of these units at first.
   6 units = $60
   Sharon had $60 at first.

Copyright © 2006 SingaporeMath.com Inc., Oregon

➤ *Challenge*: The ratio of the number of guppies to the number of angel fish to the number of swordtails in an aquarium is 28 : 9 : 21. When 24 guppies were removed, $\frac{2}{5}$ of the remaining fish were guppies. How many swordtails were there? (63)

Solution:

If $\frac{2}{5}$ of the remaining fish are guppies, then $\frac{3}{5}$ of the remaining fish are the angelfish and swordtails. Their quantities do not change. So the number of units that is the sum of the angelfish and swordtails will be $\frac{3}{5}$ of the remaining fish.

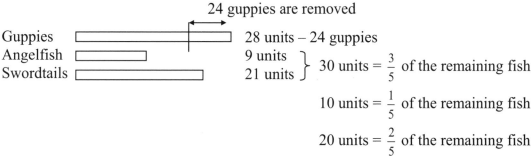

24 guppies are removed

Guppies   28 units – 24 guppies
Angelfish   9 units
Swordtails   21 units   } 30 units = $\frac{3}{5}$ of the remaining fish

10 units = $\frac{1}{5}$ of the remaining fish

20 units = $\frac{2}{5}$ of the remaining fish

20 units is $\frac{2}{5}$ of the remaining fish, and are how many units of guppies remain. Therefore,

8 units were removed. 8 units = 24 fish; 1 unit = $\frac{24}{8}$ = 3 fish

There are 21 units of swordtails; 21 units = $3 \times 21 = 63$ fish.
There were 63 swordtails.

➤ *Challenge*: John is $\frac{1}{3}$ Sam's age. In 30 years, the ratio of John's age to Sam's age will be 3 : 4. How old will Sam be then? (48 years)

Solution:

A key to this problem is that the difference between their ages stays the same. The difference now is 2 units. In 30 years, the difference will be 1 unit. So 2 units now is the same as 1 unit in 30 years from now. We can see from the diagram that 30 years is the same as 5 of the "before" units.

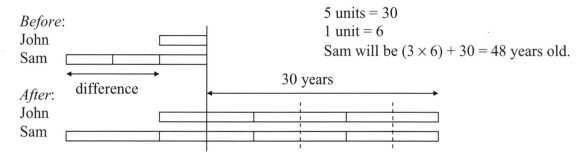

Before:
John
Sam

5 units = 30
1 unit = 6
Sam will be $(3 \times 6) + 30 = 48$ years old.

difference   30 years

After:
John
Sam

Workbook Exercise 14

Copyright © 2006 SingaporeMath.com Inc., Oregon

Review

Objectives

• Review previous material.

Suggested number of sessions: 6

	Objectives	Textbook	Workbook	Activities
Review				**6 sessions**
22-29	▪ Review	pp. 39-42, Review A pp. 43-46, Review B	Review 1	

Notes

Reviews in *Primary Mathematics* cover material from all previous levels, so the reviews are lengthy. If your students are new to *Primary Mathematics* and have difficulty with some of the topics in the reviews, you will need to re-teach some of the topics. The most effective way to do this (unless you are quite familiar both with the material and the way it was covered in earlier levels), is to go back to the levels in which it was covered. Topics that may cause problems are:

• Fractions
 3B unit 6 covers equivalent fractions and ordering fractions.
 4A unit 3 covers addition and subtraction of related fractions (fractions where the denominators are the same, or one denominator is a simple multiple of the other); mixed numbers and improper fractions; and multiplying a fraction by a whole number.
 5A unit 3 covers addition and subtraction of unrelated fractions (fractions where the denominator of one is not a simple multiple of the denominator of the other); mixed numbers, multiplication of fractions; and division of a fraction by a whole number

• Area and volume
 4A unit 7 covers area and perimeter of composite figures.
 4B unit 6 covers volume.
 5A unit 4 covers area of a triangle.
 5B unit 9 covers finding an unknown edge of a cuboid when given the volume and the other two edges or the area; and finding the volume of irregular objects by displacement of water.

• Order of operations
 5A unit 1 part 6

• Unknown angles
 5A unit 6, 5B unit 6

• Average:
 5B unit 3

• Rate
 5B unit 4

• Word problems
 3A unit 2 part 1 and part 4, unit 3 part 2
 4A unit 3 part 5
 5A unit 1 part 7, unit 3 part 7

Copyright © 2006 SingaporeMath.com Inc., Oregon

The reviews in the text tend to be a bit more challenging than the reviews in the workbook, and the problems can be good opportunities for discussion of concepts and solutions.

There are usually several ways to solve a problem. Solutions to selected problems are given here, but a student may come up with an equally valid approach. Comparing and discussing different solutions can help students' understanding of concepts.

Review A, pp. 39-42

12. $\dfrac{2 \text{ bags}}{15 \text{ cakes}} = \dfrac{10 \text{ bags}}{? \text{ cakes}}$ $2 \times 5 = 10$ bags, so $15 \times 5 = 75$ cakes

OR

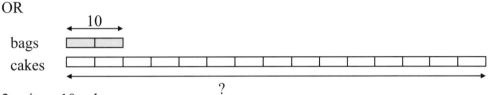

2 units = 10 cakes
1 unit = 5 cakes
15 units = $5 \times 15 = 75$ cakes

16. Total weight of 2 of the men = $54.9 \times 2 = 109.8$ kg
Weight of third man = total weight of 3 men – weight of 2 men
$= 164.4 \text{ kg} - 109.8 \text{ kg} = 54.6 \text{ kg}$

17. Total number of T-shirts sold = $200 - 20 = 180$
Total money made from sales = $180 \times \$5 = \$900$
The profit was \$360.
Cost = Sales – Profit = \$900 – \$360 = \$540
Cost price of 1 shirt = $\dfrac{540}{200} = \$2.70$

18. $\dfrac{5}{2} = \dfrac{100}{?}$ $? = 2 \times 20 = 40$ 100 oranges cost \$40 (or draw bar diagrams)
OR 5 oranges cost \$2.

1 orange costs $\$\dfrac{2}{5}$ (unitary approach)

100 oranges cost $\$\dfrac{2}{5} \times 100 = \$2 \times 20 = \$40$

Amount of money still needed = \$40 – \$35.50 = \$4.50

19. $\dfrac{1}{2}$ kg

$\dfrac{1}{2}$ of a cake weighs $2 \times \dfrac{1}{2}$ kg = 1 kg

Copyright © 2006 SingaporeMath.com Inc., Oregon

22.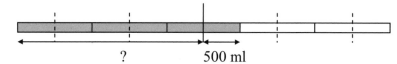

? .... 500 ml

1 unit = 500 ml
10 units = 500 × 10 = 5000 ml = 5 ℓ
The tank has a capacity of 5 ℓ

23.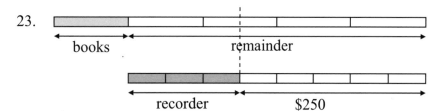

(a) Fraction of money left = $\frac{5}{8} \times \frac{4}{5} = \frac{1}{2}$

(b) Amount at first = 2 × $250 = $500

24.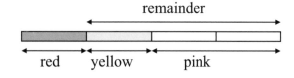

There is 1 unit more pink roses than red.
1 unit = 24
4 units = 24 × 4 = 96
There are 96 roses.

27.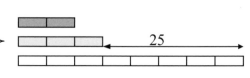

(a) swordtails : angelfish : guppies = 2 : 3 : 8

(b) There are 5 more units of guppies than of angelfish, and 13 units altogether.
    5 units = 25
    1 unit = 25 ÷ 5 = 5
    13 units = 5 × 13 = 65
    There are 65 fish.

28. John has 6 units. Peter has 3 units. Henry has 5 units.
    Henry has 2 more units than Peter.

    2 units = $80

    1 unit = $\frac{80}{2}$ = $40

    6 units = $40 × 6 = $240
    John has $240.

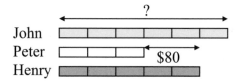

Copyright © 2006 SingaporeMath.com Inc., Oregon

29. Osman's money does not change.

Ratio of Aziz to Osman at first = 2 : 1
Use the equivalent ratio, 6 : 3
Ratio becomes 4 : 3 after Aziz spends $10.
So Aziz loses 2 units.
2 units = $10
1 unit = $10 ÷ 2 = $5
6 units = $5 × 6 = $30
3 units = $5 × 3 = $15
At first, Aziz had $30 and Osman had $15.

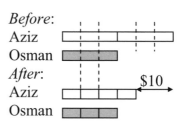

30. (a)   1 min ⟶ 80 words
             10 min ⟶ 80 × 10
                    = 800 words
             He can type 800 words in 10 min.

     (b)   80 words ⟶ 1 min

           1 word ⟶ $\frac{1}{80}$ min

           2000 words ⟶ $\frac{1}{80}$ × 2000 min = 25 min

           It will take him 25 min.

Review B, pp. 43-46

8. $\frac{3}{5}$ is smaller than $\frac{3}{4}$ (same numerator, larger denominator).

     $\frac{3}{5} = \frac{18}{30}$, $\frac{5}{6} = \frac{25}{30}$ so $\frac{3}{5}$ is smaller than $\frac{5}{6}$

     $\frac{3}{5} = \frac{9}{15}$, $\frac{2}{3} = \frac{10}{15}$ so $\frac{3}{5}$ is smaller than $\frac{2}{3}$

     $\frac{3}{5}$ is smallest.

     You can also have students find equivalent fractions with a common denominator for all
     four fractions. Then it is easy to see all at once how they compare.

     $\frac{3}{5}, \frac{2}{3}, \frac{3}{4}, \frac{5}{6}$ ($\frac{36}{60}, \frac{40}{60}, \frac{45}{60}, \frac{50}{60}$)

19. (In the 3rd edition, cherries are rambutans)
     Cost of 4 mangoes = $8.80
     Cost of 1 mango   = $8.80 ÷ 4 = $2.20
     Cost of 5 mangoes = $2.20 × 5 = $11.00
     Cost of 3 kg of cherries + 5 mangoes = $15.50
     Cost of 3 kg of cherries = $15.50 − $11.00 = $4.50
     Cost of 1 kg of cherries = $4.50 ÷ 3 = $1.50

Copyright © 2006 SingaporeMath.com Inc., Oregon

20. Total weight of 4 sacks = $18 \times 4 = 72$ kg

    Total weight of 3 sacks = $17.50 \times 3 = 52.5$ kg

    Weight of fourth sack = weight of all 4 − weight of 3 sacks = 72 kg − 52.5 kg = 19.5 kg

    OR:

    The three sacks are each 0.5 kg less than the average weight of 18 kg. So the weight of the fourth sack must be the average weight plus the weight the others are short on.

    Weight of fourth sack = $18 + (3 \times 0.5) = 19.5$ kg

22. Fraction of butter cookies $= \dfrac{1}{3} = \dfrac{5}{15} = 5$ units

    Fraction of cherry cookies $= \dfrac{2}{5} = \dfrac{6}{15} = 6$ units

    Fraction of chocolate cookies $= \dfrac{15}{15} - \dfrac{5}{15} - \dfrac{6}{15} = \dfrac{4}{15} = 4$ units

    There are $\dfrac{1}{15}$, or 1 unit more, cherry cookies than butter cookies.

    1 unit = 30 cookies

    4 units = $30 \times 4 = 120$

    There are 120 chocolate cookies.

23. 2 units = $9

    1 unit = $9 \div 2 = \$4.50$

    cost of noodles + cost of drink = $4.50

    cost of drink = $4.50 − cost of noodles

                 = $4.50 − $3.60 = $0.90

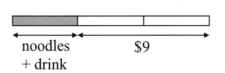

noodles
+ drink $9

24. On Monday 42 pages were read.

    Fraction of book read on Tuesday $= \dfrac{2}{5}$

    Fraction of book eft $= \dfrac{1}{4}$

    Fraction read on Monday $= 1 - \dfrac{2}{5} - \dfrac{1}{4} = \dfrac{20}{20} - \dfrac{8}{20} - \dfrac{5}{20} = \dfrac{7}{20} = 7$ units.

    7 units = 42 pages

    1 unit $= \dfrac{42}{7} = 6$ pages

    20 units = $6 \times 20 = 120$ pages

    There are 120 pages in the book.

Copyright © 2006 SingaporeMath.com Inc., Oregon

25. US

Ryan
Juan
$500
Raju
Samy
☐ = 1 unit

3d

(a) Juan (Samy) has 5 units; Ryan (Raju) started with 3 units + $80.
8 units = $500 − $80 = $420
1 unit = $420 ÷ 8 = $52.50
5 units = $52.50 × 5 = $262.50
Juan (Samy) has $262.50

(b) 3 units + $80 = (3 × $52.50) + $80 = $157.50 + $80 = $237.50
OR: $500 − $262.50 = $237.50
Ryan (Raju) had $237.50 at first.

26. 4 units = 24
1 unit = 24 ÷ 4 = 6
He used 6 cups milk.

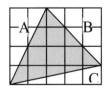

flour
milk
24

30. children

adults

men   women

men : women : children = 5 : 3 : 4

31. *Before*
Ali
David

After
Ali
David

(a) Ali's money : David's money = 3 : 5

(b) If David gives half his money to Ali, he gives $2\frac{1}{2}$ of the original units.

Ali's money : David's money = $5\frac{1}{2} : 2\frac{1}{2} = 11 : 5$

32. Area of square = 5 × 4 = 20 cm^2

Area of A = $\frac{1}{2}$ × 2 × 4 = 4 cm^2

Area of B = $\frac{1}{2}$ × 3 × 3 = $4\frac{1}{2}$ cm^2

Area of C = $\frac{1}{2}$ × 5 × 1 = $2\frac{1}{2}$ cm^2

A   B

C

Shaded area = area of square − area of A − area of B − area of C

= 20 − 4 − $4\frac{1}{2}$ − $2\frac{1}{2}$ = 9 cm^2

Copyright © 2006 SingaporeMath.com Inc., Oregon

34.   Perimeter of P = (2 × length of P) + (2 × width of P)

$$32 = 2 \times 12 + (2 \times \text{width of P})$$
$$32 = 24 + (2 \times \text{width of P})$$
$$32 - 24 = (2 \times \text{width of P}) = 8$$
$$\text{width of P} = \frac{1}{2} \times 8 = 4 \text{ cm}$$

Area of P = length of P × width of P = 12 cm × 4 cm = 48 cm^2 = area of Q

$$\text{Area of Q} = \text{length of Q} \times \text{width of Q}$$
$$48 = \text{length of Q} \times 6$$
$$\text{length of Q} = 48 \div 6 = 8 \text{ cm}$$

Perimeter of Q = (2 × length of Q) + (2 × width of Q) = (2 × 8) + (2 × 6) = 16 + 12 = 28 cm

Workbook Review 1

10.   7.2 ℓ = 7200 ml

400 ml → 1 min

$$1 \text{ ml} \to \frac{1}{400} \text{ min}$$

$$7200 \text{ ml} \to \frac{1}{400} \text{ min} \times 7200 = \frac{72}{4} \text{ min} = 18 \text{ min}$$

It will take 18 min for 7.2 ℓ to flow out of the tank.

13.   Fraction of children that were girl scouts = $\frac{4}{5} \times \frac{2}{3} = \frac{8}{15}$

14.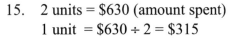

6 units were given to her mother, leaving a remainder of 2 units, one of which was given to her brother.
8 units = $80

$$1 \text{ unit} = \frac{80}{8} = \$10$$

She had $10 left.

Or:

Money left = $\frac{1}{2}$ of $\frac{1}{4}$ of 80

$$= \frac{1}{2} \times \frac{1}{4} \times 80 = \frac{80}{8} = \$10$$

15.   2 units = $630 (amount spent)
1 unit = $630 ÷ 2 = $315
3 units = $315 × 3 = $945
He saved $945.

Copyright © 2006 SingaporeMath.com Inc., Oregon

20. (a) Number of adults : total = 3 : 4

adults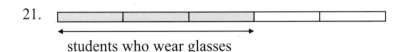
children

(b) $\dfrac{\text{number of children}}{\text{total}} = \dfrac{1}{4}$

(c) 1 unit = 207 children
4 units = 207 × 4 = 828
There are 828 people.

21.

students who wear glasses

220 + 260 = 480 (students who wear glasses)
3 units = 480
1 unit = 480 ÷ 3 = 160
5 units = 160 × 5 = 800
There are 800 students.

Or:

$\dfrac{3}{5}$ of the students = 480

$\dfrac{1}{5}$ of the students = 480 ÷ 3 = 160

All of the students = 160 × 5 = 800

24. length
width

length
width

2

(a) 1 unit = 2 cm
5 units = 5 × 2 = 10 cm
Original length was 10 cm.

(b) final length = 6 × 2 cm = 12 cm
width = 3 × 2 cm = 6 cm
Area after the increase = 12 cm × 6 cm = 72 cm²

25. Area A = $\dfrac{1}{2}$ × 5 × 3 = $7\dfrac{1}{2}$

Area B = Area C = $\dfrac{1}{2}$ × 1 × 4 = 2

Area rectangle = 5 × 4 = 20

Shaded area = area of rectangle − area of 3 triangles = $20 - 7\dfrac{1}{2} - 2 - 2 = 8\dfrac{1}{2}$ cm²

26. Height of top triangle = 18 − 10 = 8 cm

Area of top triangle = $\dfrac{1}{2}$ × 10 × 8 = 40 cm²

Area of left triangle = $\dfrac{1}{2}$ × 6 × 10 = 30 cm²

Area of square = 10 × 10 = 100 cm²
Total area = 100 + 40 + 30 = 170 cm²

Copyright © 2006 SingaporeMath.com Inc., Oregon

27. Find the cost of magazines in terms of the cost of comics.
     2 magazines → 3 comics

     1 magazine → $\frac{3}{2}$ comics

     12 magazines → $\frac{3}{2} \times 12 = 18$ comics

     27 comics + 12 magazines cost \$126 $\longrightarrow$ 27 comics + 18 comics cost \$126
     45 comics cost \$126; 1 comic costs \$126 ÷ 45 = \$2.80

     OR: If 3 comics cost the same as 2 magazines, find out how many comics he could have
     bought instead of the magazines. For 2 magazines, he could have bought 3 comics. So for
     12 magazines, how many comic books could he buy? Set up a proportion:
     $\frac{3 \text{ comics}}{2 \text{ magazines}} = \frac{? \text{ comic}}{12 \text{ magazines}}$. He could buy 18 comics for the same money as 12 magazines. If

     he just bought comics, he would have bought 27 + 18 = 45 comics.
     45 comics cost \$126; 1 comic costs \$126 ÷ 45 = \$2.80

     OR: Make equivalent units that represent 3 comics and also 2 magazines. Then, 27 comics
     is 9 units (27 ÷ 3 = 9) and 12 magazines is 6 units (12 ÷ 2).
     9 units + 6 units = 15 units. 15 units cost \$126
     1 unit costs \$126 ÷ 15 = \$8.40
     Since there are 3 comics in a unit then 3 comics cost \$8.40.
     1 comic costs \$8.40 ÷ 3 = \$2.80

28.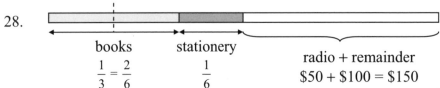

     books   stationery
     $\frac{1}{3} = \frac{2}{6}$ $\frac{1}{6}$

     radio + remainder
     \$50 + \$100 = \$150

     He spent half his money on books and stationery $\left(\frac{2}{6} + \frac{1}{6} = \frac{3}{6} = \frac{1}{2} \right)$. The other half of his

     money is the \$50 spent for the radio plus the \$100 he still had left.
     All his money = 2 × \$150 = \$300

29.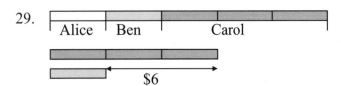

     Alice   Ben   Carol

     \$6

     2 units = \$6
     1 unit = \$6 ÷ 2 = \$3
     Ben received \$3.

30. Average weight = 3.2 kg = 3200 g
     Total weight = 3200 × 3 = 9600 g
     If another 400 g is added, total
     weight would be 4 units.
     4 units = 9600 g + 400 g = 10,000 g

     1 unit = $\frac{10,000}{4}$ g ÷ 4 = 2500 g

     A
     B   3.2 kg × 3
     C
        400 g

     Weight of C = 1 unit − 400 g = 2500 g − 400 g = 2100 g = 2 kg 100 g (or 2.1 kg)

Copyright © 2006 SingaporeMath.com Inc., Oregon

Unit 4 – Percentage

Objectives

- Express part of a whole as a fraction or as a percentage.
- Relate percentage to fractions and to decimals.
- Express one quantity as a percentage of another.
- Find the whole or a percentage part when given the value of a percentage part.
- Solve word problems which involve percentage.

Suggested number of sessions: 18

	Objectives	Textbook	Workbook	Activities
Part 1 : Part of a Whole as a Percentage				**8 sessions**
30	▪ Express part of a whole as a fraction or as a percentage.	pp. 48-48, tasks 1-3	Ex. 15, #1-2	4.1a
31	▪ Express a fraction as a percentage.	p. 47 pp. 49-50, tasks 3-6	Ex. 15, #3	4.1b
32	▪ Express a percentage as a fraction in its simplest form.	p. 50, task 7 p. 53, Practice 4A, 1	Ex. 16, #1	4.1c
33	▪ Express a decimal of up to 3 decimal places as a percentage ▪ Express a percentage as a decimal.	p. 50, tasks 8-11 p. 53, Practice 4A, 2-3	Ex. 16, #2-3	4.1d
34	▪ Solve word problems which involve percentage.	p. 51, task 12 p. 53, Practice 4A, 4-9	Ex. 17	4.1e
35	▪ Find the percentage of a percent.	p. 51, task 13 p. 53, Practice 4A, 10	Ex. 18	4.1f
36	▪ Review the meaning of tax, percent increase, percent decrease, interest, and discount. ▪ Solve word problems which involve percentage.	p. 52, tasks 14-16	Ex. 19	4.1g
37	▪ Practice.	p. 54, Practice 4B		4.1h
Part 2 : One Quantity as a Percentage of Another				**5 sessions**
38	▪ Express one quantity as a percentage of another, using the fraction method.	p. 57, tasks 1-3	Ex. 20	4.2a
39	▪ Express one quantity as a percentage of another, using the unitary method.	p. 57, tasks 1-3 pp. 55-56	Ex. 20	4.2b
40	▪ Find percent increase, percent decrease, discounts. ▪ Relate selling price to cost price as percentage.	pp. 57-58, tasks 4-6	Ex. 21	4.2c

	Objectives	Textbook	Workbook	Activities
41	▪ Find amounts when given their increase or decrease in percentage.	pp. 58-59, tasks 7-9	Ex. 22	4.2d
42	▪ Practice.	p. 60, Practice 4C	Ex. 23	4.2e
Part 3 : Solving Percentage Problems by Unitary Method				**5 sessions**
43	▪ Find the whole when given the value of a percentage part, using a unitary method.	p. 62, tasks 1-2 p. 67, Practice 4D, 1-4	Ex. 24	4.3a
44	▪ Find an original value when given a new value after its percent increase or decrease. ▪ Find a new value when given the original value and a percent increase or decrease.	p. 61 pp. 63-64, tasks 3-6 p. 67, Practice 4D, 5-8	Ex. 25	4.3b
45	▪ Solve problems which involve percentage by using the unitary method.	pp. 65-66, tasks 7-10 p. 67, Practice 4D, 9-10	Ex. 26	4.3c
46 47	▪ Practice.	p. 67, Practice 4D, 4E		4.3d

Copyright © 2006 SingaporeMath.com Inc., Oregon

| **Part 1: Part of a Whole as a Percentage** | **8 sessions** |

Objectives

- Express a part of a whole as a percentage.
- Express a percentage as a fraction in its simplest form.
- Express a percentage as a decimal.
- Solve word problems which involve percentage.

Materials

- Fraction and percentage circles (circles showing 100 divisions around the circumference)

Homework

- Workbook Exercise 15
- Workbook Exercise 16
- Workbook Exercise 17
- Workbook Exercise 18
- Workbook Exercise 19

Notes

In *Primary Mathematics 5B*, students learned to express a part of a whole as a percentage. This is reviewed in this section. If additional review would be useful, refer to unit 2, part 1 of the *Primary Mathematics 5B* textbook and the accompanying teacher's guide.

Fractions are parts of a whole. A percentage is simply a specific type of fraction with a denominator of 100. Instead of writing it as some number over 100, we write it as that number followed by a percent sign. So 1 percent, which is written as 1%, is one part out of a hundred, or $\frac{1}{100}$. 55% means 55 out of 100, or $\frac{55}{100}$.

We can convert a fraction into a percentage by finding its equivalent fraction with a denominator of 100. $\frac{11}{25}$ is 11 out of 25. By multiplying both the numerator and denominator by 4, we convert it into its equivalent fraction with a denominator of 100:

$$\frac{11}{25} = \frac{44}{100} = 44\%.$$

We can also convert a fraction to a percentage by thinking of it as a fraction of the whole, with the whole being 100%. 1 = 100%.

$$\frac{11}{25} \text{ of the whole} = \frac{11}{25} \text{ of } 100\% = \frac{11}{25} \times 100\% = \frac{11 \times 100}{25}\% = \frac{11 \times \cancel{100}^{4}}{\cancel{25}_{1}} = 44\%$$

This method is preferable when it is not easy to quickly find an equivalent fraction with a denominator of 100.

A third method for converting a fraction into a percentage is to convert the fraction into a decimal first.

$$\frac{11}{25} = 0.44 = 44\%$$

This method, which involves more computation ($11 \div 25$), is not used in *Primary Mathematics*.

In *Primary Mathematics 5B*, students learned also to express a percentage as a fraction in its simplest form, or to express a 1-place or 2-place decimal as a percentage. This is reviewed here, and extended to 3-place decimals

Since a percentage is the number of parts out of 100, we can convert a percentage into a decimal by simply moving the decimal point over two places to the left. We are dividing the percent's number by 100.

$$25\% = 0.25 \qquad\qquad 125\% = 1.25$$

To convert a decimal to a percentage, we multiply the decimal by 100, since it is a fraction *of* 100%. This simply involves moving the decimal two places to the right.

$$0.175 = 0.175 \times 100\% = 17.5\%$$

To convert a whole number percentage to a fraction in its simplest form, we first express the percentage as a fraction with a denominator of 100, and then we simplify the fraction.

$$25\% = \frac{25}{100} = \frac{1}{4} \qquad\qquad 125\% = \frac{125}{100} = 1\frac{1}{4} \ \text{ or } 125\% = 100\% + 25\% = 1\frac{25}{100} = 1\frac{1}{4}$$

Students will not encounter decimal percentages or fractions which give non-terminating decimals in this unit.

This section will also review word problems, including those which involve percent increase or decrease. One new concept will be added, that of finding a percent of a percent.

To find 25% of 60%, we convert 25% into a fraction and find it as a fraction of 60%. So 25% of 60% is $\frac{25}{100} \times 60\% = 15\%$.

Since students will start encountering more complicated percentage problems in the next few sections, it will become increasingly important to determine clearly what is to be taken as the whole quantity (the base). Use the simpler problems in this section to emphasize the quantity that is the whole.

Copyright © 2006 SingaporeMath.com Inc., Oregon

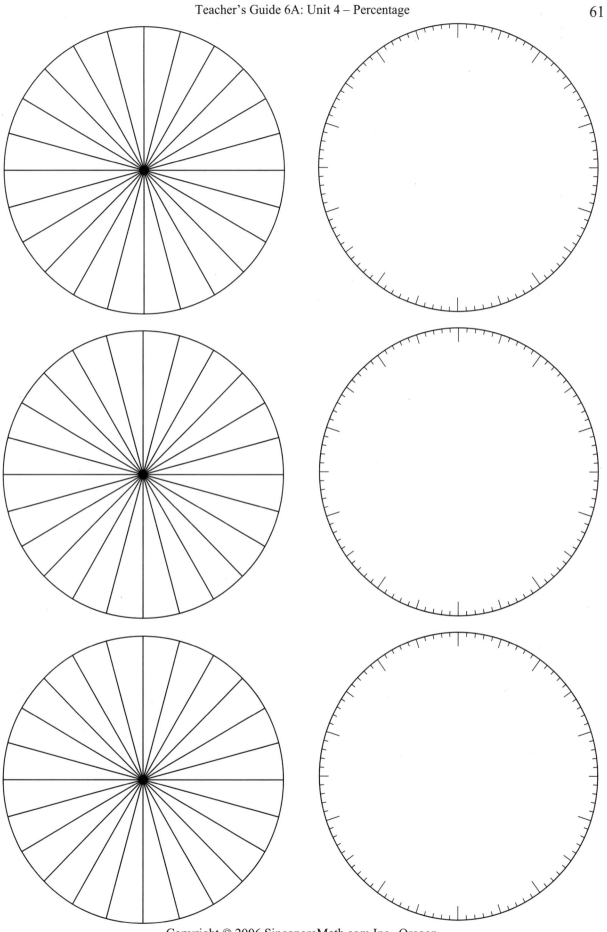

Copyright © 2006 SingaporeMath.com Inc., Oregon

Activity 4.1a   **Percentage**

1. Review percentage.
   - See if students remember what a percentage is. Remind them that a percentage is the number of parts out of 100. The whole has been divided into 100 parts, and a percent is a specific number of those parts. Instead of writing this as a fraction with a denominator of 100, we write it as the number of parts followed by the percent symbol, %. For example, 10 parts out of 100, or $\dfrac{10}{100}$, is 10%.
   - Briefly discuss ways percentages are used; e.g., the percent score on a test, the percent humidity, the percent chance of rain.

2. Discuss converting part of a whole, or a fraction, to a percentage by finding an equivalent fraction with a denominator of 100.
   - Discuss **task 1(a), textbook p. 48**.
     - The larger rectangle is the whole. On the left, it is drawn with 13 out of 50 squares shaded. The whole can be divided into 100 parts instead. This is shown on the right, with 26 smaller units shaded. Each smaller unit is half the size of the original units, so that the larger square is now divided into 100 parts. The new fraction is an equivalent fraction of the original one. $\dfrac{13}{50} = \dfrac{26}{100} = 26\%$
   - Provide students with fraction circle charts and percentage charts. You can copy the ones on p. 61 of this guide. Ask them to shade a particular fraction and then find the percentage shaded. For example, have them divide a fraction circle into fifths, shade $\dfrac{3}{5}$, and then find the percentage. They can cut out the percentage circle, which is divided into hundredths, and lay it over the fraction circle to determine the percentage. Have them also write equivalent fractions: $\dfrac{3}{5} = \dfrac{60}{100} = 60\%$.
   - Discuss **task 1(b), textbook p. 48**.
     - The larger rectangle again is the whole. In the top figure, 120 out of 300 parts are shaded. To divide the whole into 100 parts instead, each new part would be three times the size of the old part, since $300 \div 3 = 100$. So $120 \div 3$, or 40 of the new parts would be shaded. Since both the denominator (total parts) and numerator (number of parts shaded) are divided by the same number, we are finding an equivalent fraction with a denominator of 100. This can be written as a percentage. $\dfrac{120}{300} = \dfrac{40}{100} = 40\%$
   - Provide other examples.
   - You may want to point out that we are finding a proportion.
     $$\dfrac{120}{300} = \dfrac{?}{100} \quad \text{or} \quad 120 : 300 = ? : 100$$

Copyright © 2006 SingaporeMath.com Inc., Oregon

3. Have students do **tasks 2-3, textbook pp. 48-49** by finding equivalent fractions with a denominator of 100.
   - For some, students will have to first simplify the fraction. For example, for task 3(f),

   $$\frac{\cancel{15}^{3}}{\cancel{250}_{50}} = \frac{6}{100} = 6\% \ .$$

4. Discuss percentages that are greater than 100%.
   - Remind students that 1 whole is 100%. Draw a bar and label it with 0%, 50%, and 100%. Shade the whole bar. Tell students that it has 100 parts, and each part is 1%. Draw another bar under it and shade the whole bar. Ask students how we could express 2 wholes as a percentage. Since 1 whole is 100 parts of 100, then 2 wholes is 200 parts of 100, or 200%.
   - Erase half of the shading in the second bar and tell students that you now have $1\frac{1}{2}$ wholes. Ask them for the percentage. (150%).
   - Discuss instances where percentages greater than 100% might be used. For example, if a population doubles, it is now 200% of its original amount.
   - Ask students to convert some mixed numbers and improper fractions to percentages. Use ones where it is easy to find an equivalent fraction with denominator of 100.

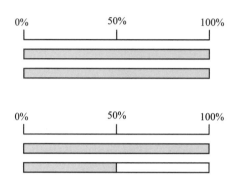

$$1\frac{3}{4} = 175\%$$

$$\frac{180}{100} = 180\%$$

$$\frac{300}{200} = 150\%$$

Workbook Exercise 15, problems 1-2

Activity 4.1b   **Fraction to percentage**

1. Discuss converting a fraction into a percentage by finding a fraction of 100%.
   - Discuss **task 4, textbook p. 49.**
     - $\frac{3}{5}$ is shown as a shaded part of a fraction bar, then as a fraction on a number line marked from 0 to 1. The fraction bar is a whole. Since 100% = 1 whole, we can draw a "percent ruler" for the same length as for the whole fraction bar.
     - Remind students that a fraction *of* a whole number is the same as the fraction *times* the number. So $\frac{3}{5}$ of 100% $= \frac{3}{5} \times 100\%$.
     - For mental math purposes, it is often easier to simplify the fraction first and then multiply, rather than the other way around.

   $$\frac{3}{5} \text{ of the whole} = \frac{3}{5} \text{ of } 100\%$$

   $$= \frac{3}{5} \times 100\%$$

   $$= \frac{3 \times 100}{5}\%$$

   $$= \frac{3 \times \cancel{100}^{20}}{\cancel{5}_{1}} = 60\%$$

Copyright © 2006 SingaporeMath.com Inc., Oregon

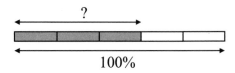

- Optional: You can relate this process to unit bar diagrams that students have been using for fractions, ratios, and proportion. The whole bar is 100%. We can find $\frac{3}{5}$ of 100% by dividing the bar into fifths, or 5 units, first finding the value of $\frac{1}{5}$, or 1 unit, and then finding the value of 3 units (60%).

5 units = 100%

1 unit $= \dfrac{100}{5}\%$

3 units $= 3 \times \dfrac{100}{5} = \dfrac{3}{5} \times 100 = 60\%$

- Have students list other fifths as percentages as well. If they know $\frac{1}{5} = 20\%$, the others are easy to compute.

$\dfrac{1}{5} = 20\%$ $\qquad$ $\dfrac{2}{5} = 40\%$

$\dfrac{3}{5} = 60\%$ $\qquad$ $\dfrac{4}{5} = 80\%$

$\dfrac{5}{5} = 100\%$

- Discuss **task 5, textbook p. 49**. Again, we can express $\frac{1}{8}$ as a percentage by finding $\frac{1}{8}$ of 100%.
- Point out that percentages can be decimals or mixed numbers. Half of a percent is half of 1 part out of 100 parts (or one part out of 200).
- Have students list other eighths as percentages as well.
  - Students should be able to easily find eighths and fourths as percentages.
  - Point out that if they memorize the percentage equivalent for $\frac{1}{4}$, they can easily find the others. $\frac{1}{8}$ is half of $\frac{1}{4}$, or 12.5%.
  - If they forget that $\frac{1}{4}$ is 25%, they can still easily calculate $\frac{1}{4}$ of 100% by finding half of 100 twice.

$\dfrac{1}{8} \times 100\% = \dfrac{1}{\cancel{8}_2} \times \cancel{100}^{25}\%$

$= \dfrac{25}{2}\%$

$= 12.5\% \text{ or } 12\tfrac{1}{2}\%$

$\dfrac{1}{8} = 12.5\%$ $\qquad$ $\dfrac{2}{8} = \dfrac{1}{4} = 25\%$

$\dfrac{3}{8} = 37.5\%$ $\qquad$ $\dfrac{4}{8} = \dfrac{2}{4} = \dfrac{1}{2} = 50\%$

$\dfrac{5}{8} = 62.5\%$ $\qquad$ $\dfrac{6}{8} = \dfrac{3}{4} = 75\%$

$\dfrac{7}{8} = 87.5\%$ $\qquad$ $\dfrac{8}{8} = 100\%$

$\dfrac{5}{8} = \dfrac{1}{2} + \dfrac{1}{8} = 50\% + 12.5\% = 62.5\%$

2. Discuss **textbook p. 47**. This page shows both methods for finding the percentage from a fraction.

3. Have students do **task 6, textbook p. 50**, using the method of finding the fraction of 100%.
   - Provide some other problems, such as those on the right, where it is easier to find the answer as a fraction of 100% rather than first changing the denominator to 100.

$\dfrac{3}{150} \times 100\% = 2\%$

$\dfrac{90}{120} \times 100\% = 75\%$

$\dfrac{18}{250} \times 100\% = 7.2\%$

Copyright © 2006 SingaporeMath.com Inc., Oregon

4. Provide some problems, such as those on the right, using fractions greater than 1.

$$2\frac{3}{5} \times 100\% = 200\% + \frac{3}{5} \times 100\% = 260\%$$

$$\frac{45}{15} \times 100\% = 300\%$$

$$\frac{78}{12} \times 100\% = 650\%$$

Workbook Exercise 15, problem 3

___

Activity 4.1c   **Percentage and fraction**

1. Discuss expressing a percentage as a fraction in its simplest form.
    *   Write a percentage, such as 25%.
    *   Remind students that a percentage gives the number of parts out of 100, so we can write it as a fraction by simply putting it over 100.
    *   Tell students to always find the fraction in its simplest form in their answers.
    *   Have students do **task 7, textbook p. 50**.
    *   Have students express some percentages greater than 100% as a fraction. These can be split into two parts, hundreds and the rest. Or students can put the whole number over 100, simplify, and change the resulting improper fraction to a mixed number.

$$25\% = \frac{25}{100} = \frac{1}{4}$$

$$125\% = 1\frac{1}{4}$$

$$308\% = 3\frac{2}{25}$$

$$235\% = 2\frac{7}{20}$$

2. Have students do **problem 1, Practice 4A, textbook p. 53**.

Workbook Exercise 16, problem 1

___

Activity 4.1d   **Percentage and decimal**

1. Discuss expressing a decimal number as a percentage.
    *   Write the decimal number 0.5 on the board. Tell students that we have two ways of converting 0.5 to a percentage.
        o   Since 0.5 is $\frac{50}{100}$, or 50 out of 100, we first convert the decimal to a fraction, and then that fraction to a percentage.
        o   We can also convert 0.5 to a percentage by finding 0.5 *of* 100 parts, or 0.5 *of* 100%. So we are multiplying 0.5 by 100. To do this, we simply move the decimal point over two places to the right.
    *   Have students express some decimals greater than 1 as percentages.

$$0.5 = \frac{50}{100} = 50\%$$

$$0.5 \times 100\% = 50\%$$

$$1.5 = 150\%$$
$$3.2 = 320\%$$
$$2.05 = 205\%$$

2. Discuss **tasks 8-9, textbook p. 50**.

Copyright © 2006 SingaporeMath.com Inc., Oregon

3. Have students do **task 10, textbook p. 50.**

4. Discuss expressing a percentage as a decimal number.
    *   Write a percentage, such as 4%, on the board.
    *   Ask students to write it as a fraction, and then as a decimal.
    *   Point out that we are dividing 4 by 100, so the result is the same as moving the decimal point two places to the left.
    *   Have students also find 40% as a decimal. Remind students 0.40 is the same as 0.4.
    *   Remind them it is customary to write zero before the decimal point, when a decimal has a value which is less than 1. That is, we write 0.04, not .04 and write 0.4 rather than .4.
    *   Have students express some percentages greater than 100% as decimals.

$$4\% = \frac{4}{100} = 0.04$$

$$40\% = 0.40 = 0.4$$

$$140\% = 1.4$$
$$115\% = 1.15$$

5. Have students do **task 11, textbook p. 50.**

6. Have students do **problems 2-3, Practice 4A, textbook p. 53.**

    Workbook Exercise 16, problems 2-3

Activity 4.1e   **Word problems**

1. Discuss **task 12, textbook p. 51**.
    *   In (a), ask students what is the whole in this problem. It is the number of students. There are 40 students, and 40 is 100% of the students.
    *   Draw a bar to illustrate this problem, and label the total as 40.
        o   Get students to help you estimate how much to shade to show 28. It is more than half the bar (20) and less than three quarters (30).
        o   Have them estimate the percentage (more than 50% of the whole bar, less than 75%, and closer to 75% than to 50%, since 28 is closer to 30 than to 20).
        o   Then have them find the percentage of that fraction of the bar which is shaded, that is, convert the fraction $\frac{28}{40}$ to a percentage. Point out that the whole, 40 students, is the denominator (bottom) in the fraction.

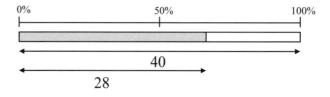

$$\frac{\overset{7}{\cancel{28}}}{\underset{1}{\cancel{40}}} \times 10\cancel{0}\% = 70\%$$

Copyright © 2006 SingaporeMath.com Inc., Oregon

2. Have students **do problems 4-8, Practice 4A, textbook p. 53** and share their solutions.
    - You can have students draw bar diagrams for these problems. Although students may not need the bar diagrams to solve these easy problems, the diagrams are a good way to organize the information and will become more important in later problems.
    - In each of the problems, make sure students pay careful attention to the wording which tells them what quantity is the whole. Call on students to explain how they determined what the whole is.

Workbook Exercise 17

Activity 4.1f   **Percentage of a percentage**

1. Discuss finding percentage of a percentage.
    - Draw a bar and mark off 60%. Draw another bar underneath the same length as the 60% part. Divide this new bar into four units.
    - Tell students that one of these units is one fourth, 25% of the 60%. Label 1 unit as 25%.
    - Now, we need to find out what percentage it is of the original bar. We can think of 100% as 100 1% units, so 60% consists of 60 1% units. We want to find 25% of 60. Since 25% is $\frac{25}{100}$, we need $\frac{25}{100}$ of 60. $\frac{25}{100} \times 60 = 15$, so 25% of 60% is 15 of the 1% units, or 15% of the original whole.
    - Recap: To find a percentage (A%) of a percentage (B%) in terms of the original (100%) total, we convert the A% into a fraction $\left(\frac{A}{100}\right)$ and multiply it by B%.
    - Have students find some other percentages of a percentage. You might have them first estimate the answers. They can also draw bars to visually see if the answer they find makes sense.

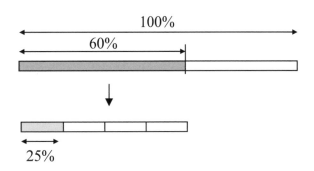

$$25\% \text{ of } 60\% = \frac{25}{100} \times 60\%$$
$$= 15\%$$

$$50\% \text{ of } 50\% = \frac{50}{100} \times 50\%$$
$$= \frac{50}{100_2} \times 50^1$$
$$= 25\%$$

$$20\% \text{ of } 80\% = 16\%$$

$$30\% \text{ of } 75\% = 22.5\%$$

2. Discuss **task 13, textbook p. 51**.

Copyright © 2006 SingaporeMath.com Inc., Oregon

3. Discuss **problem 10, Practice 4A, textbook p. 53**. Possible solution:

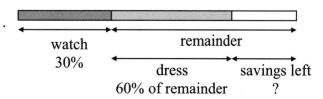

Remainder: 100% of the total – 30% of the total = 70% of total
Savings left: 100% of the remainder – 60% of the remainder = 40% of remainder

40% of 70% of the total = $\dfrac{40}{100} \times 70\% = 28\%$ of the total

She has 28% of her savings left.

Workbook Exercise 18

1. (a)

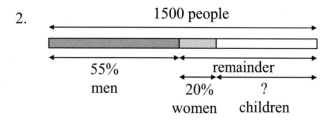

Remainder = 100% of the total – 20% of the total = 80% of the total
Percentage of money left = Remainder – 25% of remainder = 75% of the remainder

75% of 80% of the total = $\dfrac{75}{100} \times 80\% = \dfrac{3}{4} \times 80\% = 60\%$ of the total

She had 60% of her money left.
OR:
The 20% (for food) is one fifth, so there are four fifths in the remainder; 25% of those four fifths is one of the four fifths, and there are three fifths left, which is 60%.

   (b) 60% of $120 = $\dfrac{60}{100} \times \$120 = \$72$

   She had $72 left.

2.

Remainder = 100% – 55% = 45% of the total
Children = 100% of remainder – 20% of remainder = 80% of the remainder

80% of 45% = $\dfrac{80}{100} \times 45\% = 36\%$ of the total

36% of 1500 = $\dfrac{36}{100} \times 1500 = 540$

There were 540 children.

Copyright © 2006 SingaporeMath.com Inc., Oregon

Activity 4.1g   **Percent increase and decrease**

1. Review percent discounts, percent increases, interest rates, and taxes.
   - Students are probably familiar with sales discounts. Discuss "sales" and when they are likely to occur, such as 10% off for winter clothes at the end of the season.
     - Ask students what a 10% discount on $100 worth of clothing would be. 10% discount means that you are being charged 10% less than the original price.

       Discount = 10% of $100

       = $10
     - Point out that this is 10% of the initial price, so the total is the initial cost of the clothing.
     - Ask students to find the final cost. The result is the amount spent on the clothing after the discount.

       Final cost = $100 – $10

       = $90

   - Discuss situations in which there might be a percent increase. For example, a person's salary could go up 10%.
     - Ask students what a 10% increase on a salary of $100 per day would be. Point out that the initial salary of $100 is the whole.

       Increase = 10% of $100

       = $10
     - Ask students to find the final salary.

       Final salary = $100 + $10

       = $110
     - Ask students: if there is first a 10% increase on a salary of $100 per day, and then a 10% decrease, would the person end up with a salary back at $100? No, that person's salary would actually be less than what he started with. This is because the total amount that we are taking a percentage of has changed. The decrease is 10% of the new salary ($110), not of the original $100.

       Decrease = 10% of $110

       = $11

       Final salary = $110 - $11

       = $99
     - Emphasize that in order to find the correct answer; it is important to determine the whole, or total, for each step of the problem.
   - Discuss interest rates.
     - When you deposit money in the bank, the interest rate is the percentage *earned* (in one year) on the amount of money you invested. When you borrow money, the interest rate is the percentage *paid* (in one year) by you, the borrower, to the bank, on the amount borrowed. That is, at the end of one year, you, the borrower, will owe the lender the original amount borrowed, plus the interest. This original amount, whether invested or borrowed, is the base, or the whole. Emphasize the amount that is the whole, or 100%, in all of these problems.
     - Ask students what an interest rate of 5% on $400 would be. Ask them to calculate it mentally. They can find 10% of $400 easily ($40) and then find half of that ($20). 5% is half of 10%. Point out that to find 10% of any number, we just move the decimal point one place to the left.
     - Ask them to find an interest rate of 2% on $150 mentally. 1% is $1.5, so 2% is twice that, or $3. To find 1% of any number, we just move the decimal point two places to the left.
   - Discuss sales tax rates. When we buy certain goods, we have to pay a percentage tax on them, which the seller then sends to a government agency.

Copyright © 2006 SingaporeMath.com Inc., Oregon

2. Discuss **tasks 14-16, textbook p. 52**.
    *   Optional: For each of these problems, you can discuss alternate solutions. Generally, though, calculations are easier if the percentage change is found, first, then added or subtracted, as shown in the textbook. (That is, it is easier to find 3% than to find 103% of a number.)

    #14.   Since there is a 20% discount, the new cost of the items is 100% – 20% = 80% of the old cost. Final cost = $\frac{80}{100}$ × $55 = $44

    #15.   If the membership increases by 15%, the new membership is 100% + 15% = 115% of the original membership. New membership = $\frac{115}{100}$ × 140 = 161

    #16.   The new cost is 100% + 3% = 103% of the old cost.   New cost = $\frac{103}{100}$ × $600 = $618

Workbook Exercise 19

Activity 4.1h   **Practice**

1. Have students do **Practice 4B, textbook p. 54** and share their solutions. Possible solutions to problems 8-10:

    #8.   Ground beef = 1600 g
          Meatballs = 350 g
          Remainder = 1600 g – 350 g = 1250 g
          She had 20% of the remainder left.
          20% of 1250 g = $\frac{1}{5}$ × 1250 g = 250 g
          She had 250 g of the ground beef left.

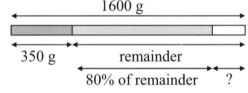

    #9.   Remainder = 100% – 15% = 85%
          85% of $80 = $\frac{85}{100}$ × $80 = $68
          Or: 15% of $80 = 10% of $80 + 5% of $80
                         = $8 + $4 = $12
          Remainder = $80 – $12 = $68

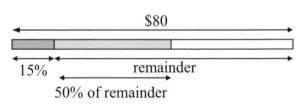

          50%, or $\frac{1}{2}$, of that is spent on meat. Amount spent on meat = $\frac{1}{2}$ × $68 = $34

    #10. Number of girls = 60% of 40 = $\frac{60}{100}$ × 40 = 24

          Number of boys = 40 – 24 = 16

          Number of girls that wear glasses = 50% of 24 = $\frac{1}{2}$ × 24 = 12

          Number of boys who wear glasses = 25% of 16 = $\frac{1}{4}$ × 16 = 4

          Total number of students who wear glasses = 12 + 4 = 16

Copyright © 2006 SingaporeMath.com Inc., Oregon

| **Part 2: One Quantity as a Percentage of Another** | **5 sessions** |

Objectives

- Express one quantity as a percentage of another.
- Solve word problems involving one quantity as a percentage of another.
- Find percent increase or decrease.
- Relate selling price to cost price as a percentage.
- Find the percentage by which one quantity is greater or less than another.
- Find a quantity when given the percentage by which it is greater or less than another.

Homework

- Workbook Exercise 20
- Workbook Exercise 21
- Workbook Exercise 22
- Workbook Exercise 23

Notes

To express a quantity as a percentage of another, the second quantity is taken as 100%. It is the base. If the first quantity is smaller than the second, the percentage is smaller than 100%. If the first quantity is greater than the second, the percentage is greater than 100%. Two methods may be used to solve these problems. For example:

Express $500 as a percentage of $400 (p. 55 in the text).

Here, $400 is the base. Note the use of the word "of" before $400. This word is the customary way to point out the quantity that is to be taken as the base. The arrows in method 2 can be read as "is equivalent to".

Method 1:
Write $500 as a fraction of $400 and then write the fraction as a percentage.

$$\frac{500}{400} \times 100\% = \frac{5}{4} \times 100\%$$
$$= 5 \times 25\%$$
$$= 125\%$$

Method 2:
Take $400 as 100%.

$$\$400 \longrightarrow 100\%$$
$$\$1 \longrightarrow \frac{100}{400}\% = \frac{1}{4}\%$$
$$\$500 \longrightarrow \frac{1}{4}\% \times 500 = 125\%$$

Express $400 as a percentage of $500.

Method 1:
Write $400 as a fraction of $500 and then write the fraction as a percentage.

$$\frac{400}{500} \times 100\% = \frac{4}{5} \times 100\%$$
$$= 4 \times 20\%$$
$$= 80\%$$

Method 2:
Take $500 as 100%.

$$\$500 \longrightarrow 100\%$$
$$\$1 \longrightarrow \frac{100}{500}\% = \frac{1}{5}\%$$
$$\$400 \longrightarrow \frac{1}{5}\% \times 400 = 80\%$$

Allow students to use either method. The first method is more common and is an extension of what students have already learned in order to convert a fraction of a quantity to a percentage, and is used in most of the examples in the text in this section. The second method, which is shown on pp. 55-56, is a unitary approach similar to what was used with rates in *Primary Mathematics 5B*. We know the relationship between one value and 100%, and we want to find the relationship between a different value and percent. By finding the percentage for 1 unit, we can find the percentage of any number of units using multiplication. If your students struggle with the second method, do not dwell on it here, particularly if they have not learned how to solve rate problems as taught in *Primary Mathematics 5B*. However, one advantage to the unitary method is that the first step involves determining the value for the base, or 100%.

When finding one quantity as a percentage of another, both quantities must be expressed in the same unit of measurement. For example, to find what percent 50¢ is of $2, we convert the $2 into cents first, and find 50¢ as a percentage of 200¢. We could also convert 50¢ to $0.50 and would get the same answer, but to avoid decimals, we generally convert the amount that represents the larger measurement unit into the smaller measurement unit.

In problems that involve a comparison between two quantities, we will be asked to find the percentage by which one quantity is more or less *than* another quantity. Note the customary usage here; the quantity that comes after the word "*than*" is the base; i.e., the quantity that we are comparing another quantity *to*. To find the percentage by which one quantity is more or less than another, first find the difference between the two quantities and then express the difference as a percentage of the smaller or larger quantity.

In the previous section of this unit, students were given an increase or a decrease as a percentage of a quantity, and asked to find the new quantity after the increase or decrease. For example, the cost of an item was given, and they were asked to find the new cost after a discount. In this section, they will be given the old and new price, and asked to find the new price as a percentage of the old price, or to find the increase or decrease as a percentage of the old price.

Finding the percentage of a quantity was taught in *Primary Mathematics 5B*. A brief review is given here before going on to find one quantity as a percentage of another. However, for those of your students who have not used *Primary Mathematics 5B*, it may be helpful to do more by going over Unit 2, Part 3 of *Primary Mathematics 5B*.

There are two methods for finding the value of a percentage-part of a whole similar to the methods for finding one quantity as a percentage of another: the fraction method, and the unitary method.

Copyright © 2006 SingaporeMath.com Inc., Oregon

For example: Find the value of 40% of 180.

Method 1:
Convert the percentage to a fraction. Then, find that fraction of the whole.

$$40\% \text{ of } 180 = \frac{40}{100} \times 180 = 72$$

Method 2:
Find the value of 1% by division. Then, multiply to find the value of other percentages.

$$100\% \text{ of } 180 = 180$$

$$1\% \text{ of } 180 = \frac{180}{100}$$

$$40\% \text{ of } 180 = \frac{18\cancel{0}}{1\cancel{0}\,\cancel{0}} \times 4\cancel{0} = 18 \times 4 = 72$$

A third method, converting the percentage to a decimal and then multiplying the decimal by the whole, is not used at this level: 40% of 180 = 0.40 × 180 = 72. This method involves more computation since there are no fractions to possibly simplify. As calculator use is not encouraged even in later levels, and mental computation is encouraged, working with fractions and simplifying them where possible can result in smaller numbers that need to be multiplied or divided, and also promotes more number-sense and use of estimation to check answers.

Activity 4.2a   **One quantity as a percentage of another**

1. Review finding one quantity as a percentage of the total, and extend this to finding one quantity as a percentage of another quantity.
   - Refer back to **task 12, textbook p. 51**.
     - 28 out of 40 students in a class walk to school. What percentage of the students walk to school?
     - We wanted to find 28 as a percentage of 40 students. We did this by finding 28 as a fraction of 40, and then multiplying that by 100% to express the fraction as a percentage: $\frac{28}{40} \times 100\%$.
   - Refer to **task 1, textbook p. 57**.
     - Here we are also asked to find a percentage of a total. What is 30¢ as a percentage of $3?
     - We can first find 30¢ as a fraction of $3. Lead students to say that both numbers have to be in the same unit of measurement in order to find this part of the total as a fraction. So first, we need to convert $3 to 300¢. Then, 30¢ is $\frac{30}{300}$ of $3.
     - To express this as a percentage, we simply multiply this fraction by 100%.

       $$\frac{\overset{1}{\cancel{30}}}{\underset{10}{\cancel{300}}} \times 100\% = 10\% \qquad \frac{\overset{10}{\cancel{30}}}{\underset{1}{\cancel{300}}} \times 1\cancel{00}\% = 10\%$$

     - Draw a bar with a percent ruler to illustrate this. Since 30¢ is about a third of a dollar, which is about a third of $3, we would shade about a third of a third as an estimate.

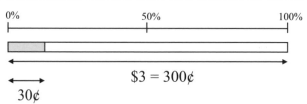

2. Discuss **tasks 2-3, textbook p. 57**.
   - Have students estimate the answer before solving, in order to be able to tell whether the answer will be less than or greater than 100%. You can have students draw rough bar diagrams for these problems. This can help them determine whether their answer makes sense. For example, in task 3, a bar for 1.35 m would be longer than one for 90 cm, so the answer should be greater than 100%.
   - Point out that in tasks 1 and 2, we converted the total to a smaller measurement unit ($3 to 300¢ and 2 ℓ to 2000 ml). Ask students why. By converting to the smaller measurement, we can avoid decimals in our calculations.
   - Point out that in task 3, the total is the smaller measure. We still convert to the smaller measure (1.35 m to 135 cm) though this time it is the part we are converting, rather than the total.

Copyright © 2006 SingaporeMath.com Inc., Oregon

3. US Edition: Provide some problems using standard U.S. measurements.
   ➢ What percentage of 5 feet is 6 inches?

       5 feet = 60 inches

       $\dfrac{6}{60} \times 100\% = 10\%$. Or, since 6 inches is half a foot, imagine the 5 feet divided into

       half-foot units. There would be 10 units, and one unit is 10%.

   ➢ 9 inches is what percentage of 5 feet?

       $\dfrac{9}{60} \times 100\% = 15\%$

   ➢ Express 1 quart as a percentage of 1 gallon.

       $\dfrac{1}{4} \times 100\% = 25\%$

   ➢ Express 1 ounce as a percentage of 1 pound.

       $\dfrac{1}{16} \times 100\% = 6.25\%$

Workbook Exercise 20

Activity 4.2b	**One quantity as a percentage of another**

1. Discuss finding one quantity as a percentage of another by using the unitary method.
   - Refer again to **task 1, textbook p. 57**. Draw a bar with a percentage ruler again and label its length as 300¢.

   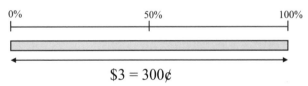

   $$\$3 = 300¢$$

   - Ask students to find 1¢ as a percentage of $3.
     - To express 1¢ as a percentage of the 300¢, we can first find it as a fraction of the total and then multiply that by 100%. 1¢ is one third of one percent of 300¢.

       $$\dfrac{1}{300} \times 100\% = \dfrac{100}{300}\% = \dfrac{1}{3}\%$$

     - Or, we can imagine the bar as divided up into 300 units. We need to find where 1 unit, or 1¢ would be on the percentage ruler. If the total is 100%, and there are 300 equal parts, how many % is in one part? We can find this by dividing 100% by the number of parts (300). So we see that 1¢ is one third of 1% on the percentage ruler. To illustrate this, draw a large percentage ruler, label 1% on it, draw a long bar under it, and show how you would start dividing it into 300 units by drawing 3 units to line up with the 1%.

       $$1¢ = \dfrac{100}{300}\% = \dfrac{1}{3}\% \text{ of 300 cents.}$$

Copyright © 2006 SingaporeMath.com Inc., Oregon

- If students have trouble with this, use simpler examples, such as the following. Draw bars to illustrate these.
  - If the total is 10, and we have two equal units, how do we find the number value of each unit? We divide: $10 \div 2 = 5$; there are 5 in each unit.
  - If the total is 10 and we have 10 equal units, how many are in each unit? We divide: $10 \div 10 = 1$. There is 1 in each unit.
  - If the total is 10 and we have 20 equal units, how many are in each unit? We divide: $10 \div 20 = \dfrac{10}{20} = \dfrac{1}{2}$. There is $\dfrac{1}{2}$ in each unit
  - If the total is 10 and we have 30 equal parts, how many are in each unit? We divide: $10 \div 30 = \dfrac{10}{30} = \dfrac{1}{3}$. There is $\dfrac{1}{3}$ in each unit.
  - If the total is 100, and we have 300 equal parts, how many are in each unit? We divide: $100 \div 300 = \dfrac{100}{300} = \dfrac{1}{3}$. There is $\dfrac{1}{3}$ in each unit.
  - You can also use a meter stick to illustrate this concept. There are 100 centimeters. If we wanted to cut the meter stick up into 300 equal sections, how long would each section be? $\dfrac{1}{3}$ of a centimeter. In the same way, if you divide a percentage ruler into 300 units, each unit would be $\dfrac{1}{3}$ of a percent.

- Tell students that since we now know what 1 cent is as a percentage of 300 cents, we can find out what any number of cents is as a percentage of 300%. Write the "arrow equations" shown here.

$$300\cent \longrightarrow 100\%$$
$$1\cent \longrightarrow \dfrac{100}{300}\% = \dfrac{1}{3}\%$$
$$30\cent \longrightarrow \dfrac{1}{3}\% \times 30 = 10\%$$

- Have students again find the solutions for the problems in **tasks 2-3, textbook p. 57**, this time using the unitary method.

$$2000 \text{ ml} \longrightarrow 100\%$$
$$1 \text{ ml} \longrightarrow \dfrac{100}{2000}\% = \dfrac{1}{20}\%$$
$$300 \text{ ml} \longrightarrow \dfrac{1}{20}\% \times 300 = 15\%$$

$$90 \text{ cm} \longrightarrow 100\%$$
$$1 \text{ cm} \longrightarrow \dfrac{100}{90}\% = \dfrac{10}{9}\%$$
$$135 \text{ cm} \longrightarrow \dfrac{10}{9}\% \times 135 = 150\%$$

2. Discuss **p. 55 in the textbook**
   - Write the problem on the board and draw the diagram as you discuss the problem, relating the information in the diagram to the information in the problem.
   - Discuss the method shown.
     - Have students tell you the quantity that is to be taken as 100%, or the base. Since we are finding Meihua's savings as a percentage **of** Sumin's savings, the quantity "Sumin's savings" is the *base*.
     - Point out that the base is the quantity pointed to by the word "of", and that this is most often the language used in these kinds of problems.

- o We can solve this by first finding what $1 is as a percentage of Sumin's savings. Once we find that percentage, we can find any quantity in dollars as a percentage of Sumin's savings.
  - o Point out that the answer will be greater than 100% because Meihua's savings is more than Sumin's savings.
- Have students also use the fraction method to solve this problem:
  - o Have them first find Meihua's savings as a fraction of Sumin's savings, then convert that fraction to a percentage by multiplying by 100%.

  $$\frac{500}{400} \times 100\% = 125\%$$

- Point out that 125% is $\frac{5}{4}$. $125\% = \frac{125}{100} = \frac{5}{4}$. Meihua's savings is $\frac{5}{4}$ of Sumin's savings.

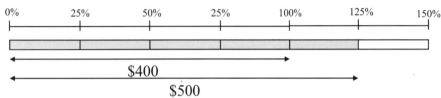

- Read the sentence in the gray box (bottom of p. 55): "Meihua saves 25% more than Sumin." Ask students how do we know what the base is, given just that phrase? This sentence could be written more explicitly as "Meihua's savings is 25% more than *Sumin's savings*." The quantity after the word *than* is the base.

3. Discuss **p. 56 in the textbook**.

- Discuss the method shown. The total ($500, Meihua's savings) is 100%. Each $1 is $\frac{100}{500}$% (or $\frac{1}{5}$%) of Meihua's savings. So Sumin's savings of $400 is $400 times the percentage for $1, or 80% of Meihua's savings.
- You can ask students to also solve this problem using fractions. That is, we first finding Sumin's savings as a fraction of Meihua's savings, and then multiply by 100%.

  $$\frac{400}{500} \times 100\% = 80\%$$

- Point out that 80% is $\frac{4}{5}$. $80\% = \frac{80}{100} = \frac{4}{5}$. Sumin's savings is $\frac{4}{5}$ of Meihua's savings.
- Read the sentence in the gray box (bottom of p. 56): "Sumin saves 20% less than Meihua." Ask students: how do we know what the base is, given just that phrase? This sentence could be written more explicitly as "Sumin's savings is 20% less than Meihua's savings." The quantity after the word **than** (less than) is the base.
- Note the thought bubble at the bottom of the page. Ask students why the two percentages are different, even though both refer to Meihua's and Sumin's savings They are different because the quantity for 100% is different. If the first quantity ($500) is 25% *more than the second* ($400), it does not follow that the second quantity ($400) is 25% *less than the first* ($500). This is an important point to make.

Workbook Exercise 20
Have students use the unitary method to solve these problems and compare to the fraction method they used earlier.

Copyright © 2006 SingaporeMath.com Inc., Oregon

Activity 4.2c   **Percent increase and decrease**

1   Discuss finding a percent difference.
   - Refer to **p. 55 in the textbook**.
   - Write the problem shown here on the board.
     o   Ask students how they would find the answer without first finding Meihua's savings as a percentage of Sumin's savings.

> Sumin saves $400 and Meihua saves $500. Find what percent more Meihua saves than Sumin does.

     o   Lead them to see that they could first find the difference in savings, and then find that as a percentage of Sumin's savings. Emphasize that we are still using Sumin's savings as the base, or 100%.

$$\text{Difference} = \$500 - \$400 = \$100$$
$$\text{Percent difference} = \frac{100}{400} \times 100\%$$
$$= 25\%$$

     o   Students can use either method to find the difference as a percentage of Sumin's savings. They have already found the percent for $1, so to use the unitary method all they need to do is multiply that by $100.
     o   Keep this problem on the board while you do the next one.
   - Refer to **p. 56 in the textbook**.
   - Write the problem shown here on the board.
     o   Again, to find the percent less that Sumin saves, we first need to find the difference ($100).

> Sumin saves $400 and Meihua saves $500. Find what percent less Sumin saves than Meihua.

     o   Then, we find this difference as a fraction of Meihua's savings, and then convert that to a percentage. Meihua's savings is now the base.
     o   Compare this equation to the previous one. Point out again that we always need to be careful to assign 100% to the correct quantity.

$$\text{Percent difference} = \frac{100}{500} \times 100\%$$
$$= 20\%$$

2   Discuss **task 4, textbook p. 57**.
   - You can ask the students to draw a rough diagram comparing cost price to selling price. The diagram also gives them an estimate of the answer. They should label one bar as 100%. Lead them to see that the base is the cost price, so that bar is labeled as 100%. The other bar is shorter, so they know their answer should be less than 100%.
   - You can ask students to also use the unitary method to solve this problem to compare to the fraction method.

$$\$1200 \longrightarrow 100\%$$
$$\$ \ \ 1 \longrightarrow \frac{100}{1200}\%$$
$$\$ \ 900 \longrightarrow \frac{100}{1200}\% \times 900 = 75\%$$

3.   Discuss **tasks 5-6, textbook p. 58**.
   - In task 5 students are finding a percentage decrease, and in task 6 they are finding a percentage increase. Make sure they understand which quantity is 100%. In task 5, it is the

Copyright © 2006 SingaporeMath.com Inc., Oregon

usual price, and in task 6 it is her weight last year. That is the value that goes in the denominator of the fraction.

Workbook Exercise 21

Activity 4.2d   **Word problems**

1. Discuss **task 7, textbook p. 58**.
    - Lead students to see that in order to find how many percent more men than women there are, we need to first find how many more men *than* women there are. Make sure students know what quantity is to be taken as the base. It is the number of women.

    Difference in number = 50 − 40 = 10

    Method 1: Express 10 as a fraction of 40 and then write that fraction as a percentage.

    $$\frac{10}{40} \times 100\% = 25\%$$

    Method 2: Take the number of women as 100%.
    40 $\longrightarrow$ 100%
    10 $\longrightarrow \frac{100}{40}\% \times 10 = 25\%$

    - Ask students to also find how many percent fewer women than men there are. The base is now the number of men (50 men).

    Method 1:
    $$\frac{10}{50} \times 100\% = 20\%$$

    Method 2:
    50 $\longrightarrow$ 100%
    10 $\longrightarrow \frac{100}{50}\% \times 10 = 20\%$

2. Review percentage of a quantity. (This review comes from pp. 33-34 in *Primary Mathematics 5B*.)
    - Write the problem given here on the board and draw a bar with a percentage ruler. Shade about 30% of it

    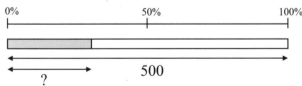

    > There were 500 people at a concert. 30% of them were children. How many children were there at the concert?

    - We want to find 30% of 500. Since percentage is one way of representing a fraction with a denominator of 100, we can find 30% of 500 by finding $\frac{30}{100}$ of 500.
    - We can still think of the total as being divided into 100 equal units. We want to find the value of 30 out of 100 of them.

    $$30\% \text{ of } 500 = \frac{30}{100} \times 500$$
    $$= 30 \times 5$$
    $$= 150$$

Copyright © 2006 SingaporeMath.com Inc., Oregon

- Tell students that we can also solve this by making our unit 1%, and first finding the value for 1%, which is 1 unit.
- If the value of 100 units is 500, then 1 unit is $500 \div 100$.
- Once we know the value of 1 unit, or 1% (5 people), we can find the value of 30 units (30%)
- Have students solve some simple problems, using both methods:
  - ➢ 5% of 300 (15)
  - ➢ 8% of 200 (16)
  - ➢ 20% of 50 (10)
  - ➢ 25% of 40 (10)
  - ➢ 45% of 70 (31.5)
  - ➢ 75% of 400 (300)

$$100\% \longrightarrow 500$$
$$1\% \longrightarrow \frac{500}{100}$$
$$30\% \longrightarrow \frac{500}{100} \times 30 = 150$$

3. Discuss **task 8, textbook p. 59**. (In the 3rd edition, Ian is Ali, and Brandon is Rahmat.)
   - Have students explain why Ian's money is the base. Since Brandon's money is being compared to Ian's money (he has 20% *more than* Ian), we take Ian's money as 100%. The 100% is one whole amount of money, which is $56.
   - Lead students to see that Brandon has 20% *more than* Ian, so he has 100% + 20% = 120% *of* (the amount of) Ian's money

4. Discuss **task 9, textbook p. 59**.
   - Have student tell you why the weight of Package A is the base. The weight of Package B (which we don't know) is being compared to the weight of Package A (5 kg), so we take the weight of Package A as 100%.
   - Package B weighs 15% *less than* Package A, so the weight of Package B is 100% − 15% = 85% of the weight *of* Package A. $\frac{85}{100} \times 5 \text{ kg} = 4\frac{1}{4} \text{ kg}$, or 4.25 kg.

Workbook Exercise 22

Copyright © 2006 SingaporeMath.com Inc., Oregon

Activity 4.2e   **Practice**

1   Have students do **Practice 4C, textbook p. 60**. Students can use either the fractional or the unitary method in solving these problems.

Workbook Exercise 23
Possible solutions:

1.   (a)

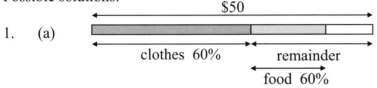

Amount spent on clothes = 60% of $50 = 6 × 10% of $50 = 6 × $5 = $30
Amount left = $50 − $30 = $20
Amount spent on food = 60% of $20 = 6 × 10% of $20 = 6 × $2 = $12
Difference between amount spent on clothes and on food = $30 − $12 = $18

(b)   We want to find the percent more spent on clothes *than* on food. The base is the amount spent on food ($12). The amount being compared to the amount spent on food is the amount more spent on clothes ($18). We are finding the *difference* as a percentage of the food amount.

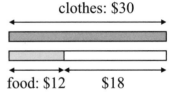

$$\frac{\text{difference}}{\text{food amount}} \times 100\% = \frac{18}{12} \times 100\% = 150\%$$

2.

(a)   John's money ($400) is 100%.

Peter's money = 120% of $400 = $\frac{120}{100} \times \$400 = \$480$

Henry's money = $480 ÷ 2 = $240
Amount they had in all = $400 + $480 + $240 = $1120

(b) $\dfrac{\text{Henry's money}}{\text{John's money}} = \dfrac{240}{400} \times 100\% = 60\%$

Copyright © 2006 SingaporeMath.com Inc., Oregon

Part 3: Solving Percentage Problems by the Unitary Method   **5 sessions**

Objectives

- Find the whole when given the value of a percentage part.
- Find the original value when given a new value and the percent by which it increased or decreased from the original value.
- Find the new value when given an original value and the percent increase or decrease from the original value.

Homework

- Workbook Exercise 24
- Workbook Exercise 25
- Workbook Exercise 26

Notes

In these problems, students are given the value of a percentage part and asked to find the whole. For example, if 75% of the whole is 12, what is the whole? Or, rephrased: 12 is 75% of what? Students will learn to find the answer by using a unitary method in which they first find the value of 1% of the whole.

$$75\% \text{ of the whole is } 12$$
$$1\% \text{ of the whole is } \frac{12}{75}$$
$$100\% \text{ of the whole is } \frac{12}{75} \times 100 = 16$$

In *Primary Mathematics 4A*, students have already learned to use a unitary approach for solving fraction problems. For example:

Mary spent $\frac{3}{4}$ of her money on a book that cost \$12. How much money did she start with?

To solve this, students drew a bar diagram, showing fourths as units, labeled 3 of them as \$12, found the value for 1 of the units, and then multiplied by 4 to find the value for all 4 units.

$$3 \text{ units } (\tfrac{3}{4} \text{ of the whole}) = \$12$$
$$1 \text{ unit } (\tfrac{1}{4} \text{ of the whole}) = \$\frac{12}{3}$$
$$4 \text{ units } (\tfrac{4}{4} \text{ of the whole}) \ \$\frac{12}{3} \times 4 = 16$$

The unitary method for solving percentage problems is very similar, except that the unit is 1% so there are 100 units on the whole bar.

$$75 \text{ units, or } 75\% \longrightarrow 12$$
$$1 \text{ unit, or } 1\% \longrightarrow \frac{12}{75}$$
$$100 \text{ units, or } 100\% \longrightarrow \frac{12}{75} \times 100 = 16$$

The arrow can be read "of the whole is" which in this case is "of her total money is".

Once we find the value for 1%, we can find the value for any other percentage-part of the whole. If we are told that 75% of the total is 12, and asked to find 25% of the total, we first find the value of 1% of the whole, and then multiply that by 25 to find the quantity which is 25% of the whole.

As students become proficient with these problems — but only then — they may skip the intermediate step:

$$75\% \longrightarrow 12 \qquad\qquad 75\% \longrightarrow 12$$
$$100\% \longrightarrow \frac{12}{75} \times 100 = 16 \qquad 25\% \longrightarrow \frac{12}{75} \times 25 = 4$$

They should be able to use further shortcuts as well to simplify the problem. In the above example, students might recognize 75% as being $\frac{3}{4}$, and find $\frac{1}{4}$, or 25% first.

$$\div 3 \left(\begin{array}{l} 75\% \longrightarrow 12 \\ 25\% \longrightarrow \frac{12}{3} = 4 \\ 100\% \longrightarrow 4 \times 4 = 16 \end{array} \right) \begin{array}{l} \div 3 \\ \\ \times 4 \end{array}$$

Some students may also convert the problem to fractions. In the example above, they can convert 75% to $\frac{3}{4}$, find the value for $\frac{1}{4}$, and then for $\frac{4}{4}$.

Note that all the examples in the text have the student finding the value for 1%. This is a standard procedure that will always work, and helps students to not lose sight of the meaning of a percent. Allow students to use shortcuts, or convert to fractions, if they come up with these approaches on their own, but encourage students to find the value for 1%, generally.

As with any of the learning tasks, you can present the problems to your students without having them first look at the text. Let students try to work out a solution and then discuss it. Then compare their approach to the text's approach.

Activity 4.3a   **Word problems**

1. Discuss **task 1, textbook p. 62**.
   - Remind students that we can think of the bar as divided up into 100 equal parts. Each part is $\frac{1}{100}$ of the total, which is 1% of the total.
   - We are told how much 75 units, or 75% of the total, is 42. We are asked to find the total score.
   - First, we find what one unit is. How? We can divide 42 by 75. This is what is shown in the first two lines of the textbook's solution. The arrow can be read as "of the total is".
   - We don't need to solve this fraction, $\frac{42}{75}$, right away. (If we did, we would find out that 1 unit is $0.56.)
   - Now that we know what 1 unit, or 1%, is, we can find 100 units, or 100%, by multiplying the value for 1% by 100.
   - We can simplify. One approach is:

   $$\frac{42}{75} \times 100 = \frac{42}{75_3} \times 100^4 = \frac{42^{14}}{75_1} \times 100^4 = 14 \times 4 = 56$$

   - Point out to your student that we do the same thing to the numbers on both side of the arrow in order to keep their relationship unchanged.

   $$\div 75 \left(\begin{array}{l} 75\% \longrightarrow 42 \\ 1\% \longrightarrow \dfrac{42}{75} \\ 100\% \longrightarrow \dfrac{42}{75} \times 100 \end{array} \right) \begin{array}{l} \div 75 \\ \\ \times 100 \end{array} \qquad \begin{array}{l} 75 : 42 = 1 : ? \\ \\ 1 : \dfrac{42}{75} = 100 : ? \end{array}$$

   $\times 100$

2. Discuss **task 2, textbook p. 62**. (In the 3rd edition, Adam is Rahim)
   - Ask students to tell you which value is the base (Adam's salary) and why (we are told that Jim's salary is 90% *of* Adam's salary).
   - The first "arrow equation" means that 90% of the total (Adam's salary) is $864. So we find the total by first finding 1%, then 100%.

3. Have students do **problems 1-4, Practice 4D, textbook p. 67** and share their answers, as time permits.

Workbook Exercise 24

#1. The total number of books is the base (100%).

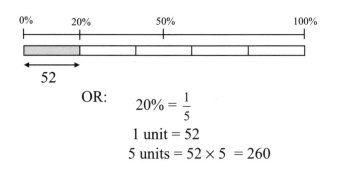

20% ⟶ 52 comic books

1% ⟶ $\frac{52}{20}$

100% ⟶ $\frac{52}{20} \times 100 = 260$ books

He has 260 books altogether.

OR:

$20\% = \frac{1}{5}$

1 unit = 52

5 units = $52 \times 5 = 260$

Copyright © 2006 SingaporeMath.com Inc., Oregon

#2.   Meifen's money is the base.

75% of Meifen's money is $300.

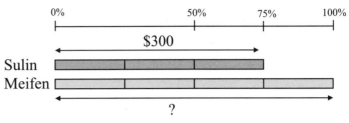

$$75\% \longrightarrow \$300$$
$$1\% \longrightarrow \$\frac{300}{75}$$
$$100\% \longrightarrow \$\frac{300}{75} \times 100 = \$400$$

Meifen saves $400.

OR:
$$75\% = \frac{3}{4}$$
3 units = $300
$$1 \text{ unit } = \$\frac{300}{3} = \$100$$
4 units = $100 × 4 = $400

#3.   Peter's total money is 100%.
Percentage spent on book = 60%
Percentage left = 100% - 60% = 40%
40% of the total money is $18.

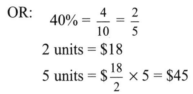

$$40\% \longrightarrow \$18$$
$$1\% \longrightarrow \$\frac{18}{40}$$
$$100\% \longrightarrow \$\frac{18}{40} \times 100 = \$45$$

Peter had $45 at first.

OR:
$$40\% = \frac{4}{10} = \frac{2}{5}$$
2 units = $18
$$5 \text{ units} = \$\frac{18}{2} \times 5 = \$45$$

#4.   The total number of stamps is 100%.
Percentage given away = 25%
Percentage left = 75%
75% of the total number of stamps left is 450.

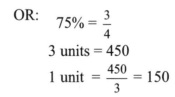

$$75\% \longrightarrow 450$$
$$1\% \longrightarrow \frac{450}{75}$$
$$25\% \longrightarrow \frac{450}{75} \times 25 = 150$$

He gave away 150 stamps.

OR:
$$75\% = \frac{3}{4}$$
3 units = 450
$$1 \text{ unit } = \frac{450}{3} = 150$$

Copyright © 2006 SingaporeMath.com Inc., Oregon

Activity 4.3b   **Word problems**

1   Discuss **the problem on p. 61 of the textbook.**
    - Ask students to tell you which value is the base and why. The increase in the selling price is being compared *to* the cost price. So the cost price is the base (100%).
    - Ask them: which quantity is given in the problem? It is the selling price ($3600).
    - Draw the textbook's illustration on the board. Tell students that if we make the cost-price bar 100%, then the selling-price bar is 100% plus another 20%, or 120%. Emphasize that the unit is 1%.
    - We are told the selling price is $3600. We can label the selling price bar with that amount. Since the selling-price bar is 120%, we can associate the selling price ($3600) with 120% of the cost price.
    - Then we can find the value for 1%, and then 100%. Or, after finding the value for 1%, we could find the value for 20%, the increase ($600) and subtract that from the selling price to get the cost price ($3600 - $600 = $3000)
    - Point out that we do not always have to go down to 1%. When the percent for which we are given the value is a multiple of 10, it is just as easy to find the value for 10%.

$$\div 12 \left(\begin{array}{l} 120\% \longrightarrow \$3600 \\ 10\% \longrightarrow \dfrac{3600}{12} \end{array} \right) \div 12$$

$$\times 10 \left(\begin{array}{l} 10\% \longrightarrow \dfrac{3600}{12} \\ 100\% \longrightarrow \dfrac{3600}{12} \times 10 \end{array} \right) \times 10$$

    - Caution students that if they use this approach, they need to keep track of what they are doing and not multiply by 100 at the end just because that might be what they usually do. They should always check if an answer makes sense. If, by mistake, they first divided by 12 and then multiplied by 100 the answer would be $30,000 instead, which doesn't make sense — for one thing, the cost price can't be more than the selling price if there was an increase.
    - Show students that we could also do this problem using fractions. Since 20% is $\frac{1}{5}$, we can use $\frac{1}{5}$ as the unit. The selling price is then 6 units, and the cost price 5 units.

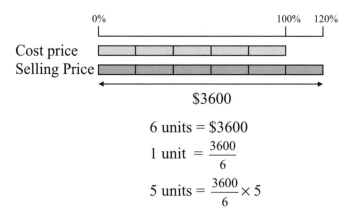

$$6 \text{ units} = \$3600$$
$$1 \text{ unit } = \frac{3600}{6}$$
$$5 \text{ units} = \frac{3600}{6} \times 5$$

2. Discuss **task 3, textbook p. 63.**
    - Have students tell you what quantity is the base and why. We are given the percent price reduction *of the usual price*, so the usual price is the base. We are given the selling price, and the discount. So we know the value for 85% ($17), and can find the value for 100%.

3. Discuss **task 4, textbook p. 63.**

Copyright © 2006 SingaporeMath.com Inc., Oregon

- We are given the percent by which the number of books in the library increased, so the original number is the base.

4. Discuss **task 5, textbook p. 64**.
   - We are given the percentage increase (10%), so the base is the monthly salary she got before the increase. We are given a value ($120) for 10% of this salary and can find the value for the salary either before the increase (100%) or after the increase (110%).
   - Since the percentage for the new salary (110%) is an easy multiple (11) of the percentage (10%) for which we have a given value, we can simply multiply by 11 as an alternate to first determining the value for 1%.

$$10\% \longrightarrow \$120$$
$$1\% \longrightarrow \$\frac{120}{10}$$
$$110\% \longrightarrow \$\frac{120}{10} \times 110 = \$1320$$

Or
$$10\% \longrightarrow \$120$$
$$110\% \longrightarrow \$120 \times 11 = \$1320$$

5. Discuss **task 6, textbook p. 64**.
   - This problem has an extra step. Lead students to see that they must first find the cost price to determine the price he must sell the watch for, and still make a profit of $150.
   - The base is the cost price. We are given a value ($600) for 80% of the cost price. From this we can find the value for 100% of the cost price. The price he must sell the watch for is $150 more than the cost price.

80% of the cost price is $600.
$$80\% \longrightarrow \$600$$
$$1\% \longrightarrow \$\frac{600}{80}$$
$$100\% \longrightarrow \$\frac{600}{80} \times 100 = \$750$$
New price: $750 + $150 = $900

6. Have students do **problems 5-8, Practice 4D, textbook p. 67** and share their solutions, as time permits. Possible solutions for problems 7 and 8:

#7.   The base is the usual price.
      Selling price = 100% − 15% = 85% of the usual price.

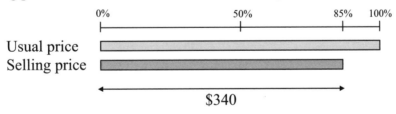

85% of the usual price is $340.
$$85\% \longrightarrow \$340$$
$$1\% \longrightarrow \$\frac{340}{85}$$
$$100\% \longrightarrow \$\frac{340}{85} \times 100 = \$400$$
The usual price of the bicycle is $400.

Copyright © 2006 SingaporeMath.com Inc., Oregon

#8.   The score for the English test is the base.
       Math score = 100% + 5% = 105% of the English test score

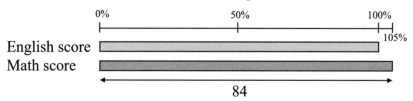

       105% of the English test score is 84.
       $105\% \longrightarrow 84$
       $1\% \longrightarrow \dfrac{84}{105}$
       $100\% \longrightarrow \dfrac{84}{105} \times 100 = 80$
       Her English test score was 80 points.

Workbook Exercise 25
Note: Calculations for 1% are shown in these solutions. Although students can skip this step, some errors may result from skipping this step. However, most errors are likely to be due to using the wrong base.

Possible solutions:

#1.   The usual price is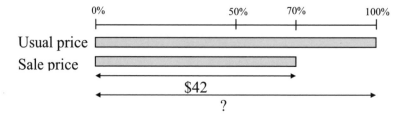
       the base (100%).
       70% of the usual
       price is $42.

       $70\% \longrightarrow \$42$ OR: $70\% = \dfrac{70}{100} = \dfrac{7}{10}$
       $1\% \longrightarrow \$\dfrac{42}{70}$ 7 units = $42
       $100\% \longrightarrow \$\dfrac{42}{70} \times 100 = \$60$ 10 units = $\$\dfrac{42}{7} \times 10 = \$60$

       The usual price is $60.

#2.   The price before the
       increase is the base.
       110% of the old
       price is $2420.
       $110\% \longrightarrow \$2420$
       $1\% \longrightarrow \$\dfrac{2420}{110}$
       $100\% \longrightarrow \$\dfrac{2420}{110} \times 100 = \$2200$
       The price before the increase was $2200.

Copyright © 2006 SingaporeMath.com Inc., Oregon

#3.   There are two separate cases. The number of boys before the change is the base for the boys, and the number of girls before the change is the base for the girls.

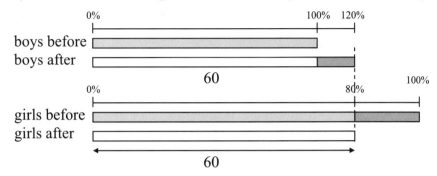

120% of the original number of boys is 60.

$120\% \longrightarrow 60$

$1\% \longrightarrow \dfrac{60}{120}$

$100\% \longrightarrow \dfrac{60}{120} \times 100 = 50$

There were originally 50 boys.

80% of the original number of girls is 60.

$80\% \longrightarrow 60$

$1\% \longrightarrow \dfrac{60}{80}$

$100\% \longrightarrow \dfrac{60}{80} \times 100 = 75$

There were originally 75 girls.

Total membership before = 50 + 75 = 125
Total membership after = 60 + 60 = 120
There was an overall decrease of 5 members.

Activity 4.3c   Word problems

1   Discuss **tasks 7-8, textbook p. 65.**
   •   Make sure in each task that students understand what value is to be taken as the base, and how to find a relationship between a percentage and a value.

2   Discuss **task 9, textbook p. 66.**
   •   We are given the percentage *of total teachers* that are male (40%), so the total number of teachers is the base.
   •   Since 40% are males, then 60% are females. The difference in percentage is 20%. This difference in the number of male and female teachers is 18. So 20% *of the total teachers* is 18.
   •   If the problem had said instead that there are 20% more female teachers *than male teachers*, then the base would be the male teachers.

20% of the total number of teachers is 18.

$20\% \longrightarrow 18$

$100\% \longrightarrow \dfrac{18}{20} \times 100 = 90$

Copyright © 2006 SingaporeMath.com Inc., Oregon

3   Discuss **task 10, textbook p. 66.**
    • In this problem, at first we can only find the value for a specific percent (rather than for
      100%). But, once we find that relationship between a percent and a number, we can find the
      value for 1%, then for any percent we want.
    • We are given the number of spaces for motorcycles, and enough information to find the
      percent of spaces that are for motorcycles and for buses. It is the number of spaces for buses
      that we have to find. So first, we will use the relationship between the *number* of spaces for
      motorcycles and the *percent* of spaces for motorcycles, and this will let us find the number of
      spaces for buses. $8\% \longrightarrow \dfrac{24}{12} \times 8 = 16$. There are 16 spaces for buses.

4   Have students do **problems 9-10, Practice 4D, textbook p. 67 and share their solutions.**
    #9. The amount he spent last week is the base.
         Amount spent this week = 100% + 10% = 110% of the amount spent last week.

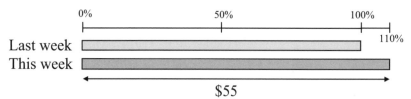

         110% of the amount spent last week is $55.
         $110\% \longrightarrow \$55$
         $100\% \longrightarrow \$\dfrac{55}{110} \times 100 = \$50$
         He spent $50 last week.

    #10. The number of girls is the base. The number of boys is 10% more than the number of girls.
         The total number of children = 100% + 110% = 210% of the number of girls. To find the
         number of children in the choir, we need to find the value for 210%.
         10% of the number of girls is 4.
         $10\% \longrightarrow 4$
         $1\% \longrightarrow \dfrac{4}{10}$
         $210\% \longrightarrow \dfrac{4}{10} \times 210 = 84$
         There are 84 children.

Workbook Exercise 26
Possible solutions:

    #1. Sunday visitors = 2300 = 115% of Saturday visitors
         115% of the number of visitors on Saturday is 2300.
         $115\% \longrightarrow 2300$
         $100\% \longrightarrow \dfrac{2300}{115} \times 100 = 2000$
         There were 2000 visitors on Saturday.

Copyright © 2006 SingaporeMath.com Inc., Oregon

#2. After buying the handbag, Mary had 60% of her money left. Shoes cost 40% of this remainder (as shown in the diagram in the workbook). The amount of money she has left after buying shoes is 60% of 60%.

Cost of shoes = 40% of 60% = $\frac{40}{100} \times 60\%$ = 24% of her money.

Money left = 100% - 40% - 24% = 36%.

Or: The amount of money she has left after buying shoes is 60% of 60%.

Money left = 60% of 60% = $\frac{60}{100} \times 60\%$ = 36%

36% of her money is $90.

36% $\longrightarrow$ $90

100% $\longrightarrow$ $\frac{90}{36} \times 100 = \$250$

She had $250 at first.

OR:

60% of the remainder is 90.

60% $\longrightarrow$ 90

100% $\longrightarrow$ $\frac{90}{60} \times 100 = \$150$

60% of her money is $150.

60% $\longrightarrow$ $150

100% $\longrightarrow$ $\frac{150}{60} \times 100 = \$250$

#3. The total number of stamps is the base. (In the 3rd edition, U.S. stamps are Singapore stamps, and Canadian stamps are Malaysian stamps.)
There are 100% – 30% = 70% U.S. stamps. The difference is 70% – 30% = 40% of the total stamps. The difference is 500.
So 40% of the total stamps is 500 stamps.

40% $\longrightarrow$ 500

100% $\longrightarrow$ $\frac{500}{40} \times 100 = 1250$

He has 1250 stamps altogether.

#4. 25% of the boys in the club is 36.

25% $\longrightarrow$ 36

100% $\longrightarrow$ $\frac{36}{25} \times 100 = 144$

There are 144 boys.

Copyright © 2006 SingaporeMath.com Inc., Oregon

Activity 4.3d   **Practice**

1   Have students do any problems in **Practice 4D, textbook p. 67** not yet done, and the problems in
    Practice 4E, textbook p. 68 and share their solutions. Some of the Practice 4E problems,
    especially #5, deserve particular attention.

Possible solutions for problems 5-10:

#5.   (In the 3rd edition, Tim is Ali, and Carlos is Osman.)
      (a)   The base is the total number of test questions (50).

            Tim answered 80% of 50 correctly: $\frac{80}{100} \times 50 = 40$ correct answers.

            Carlos answered 90% of 50 correctly: $\frac{90}{100} \times 50 = 45$ correct answers.

            Carlos answered 45 − 40 = 5 more questions correctly than Tim did.

      (b)   Determining the base is tricky in this problem. The question asks how many percent
            more questions Carlos answered correctly *than* Tim. Since the difference is being
            compared to the number of questions Tim answered correctly, then the base is the
            number of questions Tim answered correctly. It is *not* the total number of questions.

            We need to find the difference as a percentage of 40 (Tim's correct answers). Draw
            bar diagrams to illustrate this for students.

            $\frac{5}{40} \times 100\% = 12.5\%$ Carlos answered 12.5% more questions correctly than Tim.

#6.   To find the *percent more* males than females, we need to find the *number more* males than
      females, so we need both the *number* of males and the *number* of females. The number of
      females is the base for this problem. Then we can find the difference between the number
      of males and females as a percentage of the number of females. The first step is to find the
      number of males.

      Number of males = 60% of 200 = $\frac{60}{100} \times 200 = 120$

      Number of females = 200 − 120 = 80
      Difference = 120 − 80 = 40 more males than females.
      80 $\longrightarrow$ 100%   OR: $\frac{40}{80} \times 100\% = 50\%$
      40 $\longrightarrow$ 50%
      There are 50% more males than females.

#7.   40% of the beads are red and 60% are yellow. There are 20% more yellow beads than red
      beads, and we are told that there are 36 more yellow beads than red beads. So 20% of the
      beads is 36.
      20% $\longrightarrow$ 36

      100% $\longrightarrow$ $\frac{36}{20} \times 100 = 180$

      There are 180 beads.

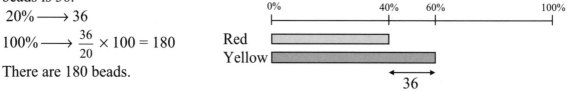

Copyright © 2006 SingaporeMath.com Inc., Oregon

#8.   Libby [Mrs Li] paid 100% − 20% = 80% of the usual price for the watch.
      80% of the usual price is $600.

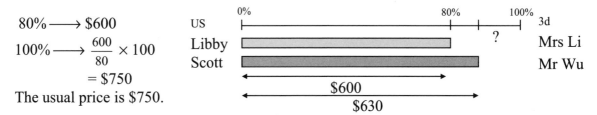

$80\% \longrightarrow \$600$

$100\% \longrightarrow \dfrac{600}{80} \times 100$

$\qquad = \$750$

The usual price is $750.

Scott [Mr Wu] paid $630 for the same watch.
So his discount was only $750 − $630 = $120.

$\dfrac{120}{750} \times 100\% = 16\%$

He was given a 16% discount.

#9.   The base is Alice's salary. Mary's salary is 110% of Alice's salary.

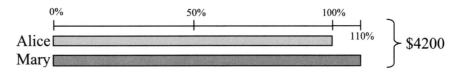

Together they have 100% + 110% = 210% of Alice's salary.

$210\% \longrightarrow \$4200$

$1\% \longrightarrow \$\dfrac{4200}{210}$

$110\% \longrightarrow \$\dfrac{4200}{210} \times 110 = \$2200$

Mary's salary is $2200.

#10. (a) After buying food, John has 80%
         of his money left.

         $\dfrac{2}{5}$ of $80\% = \dfrac{2}{5} \times 80\% = 32\%$

         He spent 32% of his money on the toy.

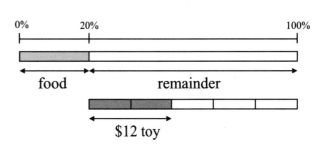

     (b) 32% of his money is $12.

         $32\% \longrightarrow \$12$

         $1\% \longrightarrow \$\dfrac{12}{32}$

         $100\% \longrightarrow \$\dfrac{12}{32} \times 100 = \$37.50$

         John had $37.50 at first.

Copyright © 2006 SingaporeMath.com Inc., Oregon

Review

Objectives

- Review previous material.

Suggested number of sessions: 5

	Objectives	Textbook	Workbook	Activities
Review				**4 sessions**
48-52	▪ Review.	pp. 69-73, Review C	Review 2	

Possible solutions to selected problems:

Review C, pp. 69-73

11.   (a) $5 \times (\underline{13 - 9}) + \underline{8 \div 4}$
                $= \underline{5 \times 4} + 2$
                $= 20 + 2$
                $= 22$

       (b) $8 \times (\underline{52 - 47}) \div 2$
                $= \underline{8 \times 5} \div 2$
                $= \underline{40 \div 2}$
                $= 20$

21. $\frac{2}{5}$ of a number is 42

     $\frac{1}{5}$ of a number is $\frac{42}{2} = 21$

     $\frac{5}{5}$ of a number is $\frac{42}{2} \times 5 = 105$

     $\frac{1}{3}$ of 105 $= \frac{1}{3} \times 105 = 35$

22.   Total of the four numbers $= 60 \times 4 = 240$
     Total of three of the numbers $= 45 + 56 + 75 = 176$
     Fourth number $= 240 - 176 = 64$

23. $\frac{1}{2}$ h $\longrightarrow$ 650 revolutions

     1 h $\longrightarrow$ $650 \times 2 = 1300$ revolutions
     3 h $\longrightarrow$ $1300 \times 3 = 3900$ revolutions

27.   1 unit $= 200 - 50 = 150$
     2 units $= 150 \times 2 = 300$
     There are 300 green balls.

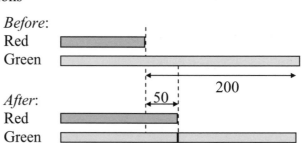

29.   5 oranges $\longrightarrow$ \$1.20   3 apples $\longrightarrow$ \$1

160 oranges $\longrightarrow$ \$$\frac{1.20}{5} \times 160 = \$38.40$ 90 apples $\longrightarrow$ \$$\frac{1}{3} \times 90 = \$30$

Total money = \$38.40 + \$30 = \$68.40

30. $\frac{1}{4}$ of Jim's savings is \$350

$\frac{4}{4}$ of Jim's savings is \$350 $\times$ 4 = \$1400

31.   How many pears can he have bought with $\frac{3}{5}$ of his money?

1 mango costs 2 times as much as a pear.
The cost of 1 mango is the cost of 2 pears.
The cost of 15 mangoes is the cost of 15×2, or 30 pears.

Buying 15 mangoes and 9 pears would cost the same as buying $30 + 9$, or 39 pears. With $\frac{3}{5}$

of his money he could have bought 39 pears.
Using this information, how many pears can he buy with the remainder of his money?

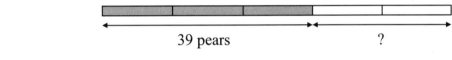

<table>
<tr><td>39 pears</td><td>?</td></tr>
</table>

$\frac{3}{5}$ of his money $\longrightarrow$ 39 pears   or: 3 units = 39

$\frac{1}{5}$ of his money $\longrightarrow$ $39 \div 3 = 13$ pears   1 unit $= \frac{39}{3} = 13$

$\frac{2}{5}$ of his money $\longrightarrow$ $13 \times 2 = 26$ pears   2 units $= 13 \times 2 = 26$

He can buy 26 pears with the rest of his money.

32.   Number of kg of grapes he sold at \$5.25 $= \frac{2}{3} \times 18 = 12$ kg

Amount received = $12 \times \$5.25 = \$63$
Remainder = $18 - 12 = 6$ kg
Amount received for the remainder = $6 \times \$4.80 = \$28.80$
Total received = \$63 + \$28.80 = \$91.80

33. $\frac{3}{5}$ full is 420 ml

$\frac{1}{5}$ full is $\frac{420}{3}$ ml

$\frac{5}{5}$ full is $\frac{420}{3} \times 5 = 700$ ml

The capacity of the bottle is 700 ml.

Copyright © 2006 SingaporeMath.com Inc., Oregon

34.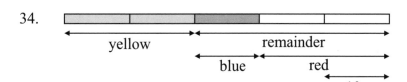

Fraction of yellow marbles = $\frac{2}{5}$ Remainder = $\frac{5}{5} - \frac{2}{5} = \frac{3}{5}$

Fraction of blue marbles = $\frac{1}{3} \times \frac{3}{5} = \frac{1}{5}$ Fraction of red marbles = $\frac{3}{5} - \frac{1}{5} = \frac{2}{5}$

Difference between red and blue marbles = $\frac{2}{5} - \frac{1}{5} = \frac{1}{5}$

$\frac{1}{5}$ of the marbles = 10   or: 1 unit = 10

$\frac{5}{5}$ of the marbles = $10 \times 5 = 50$ 5 units = $10 \times 5 = 50$

There are 50 marbles.

35.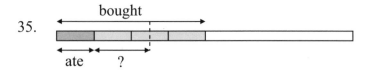

Remainder = $\frac{3}{4}$ of $\frac{1}{2}$ of cake = $\frac{3}{8}$ of cake

Each piece = $\frac{1}{2}$ of remainder = $\frac{1}{2}$ of $\frac{3}{8}$ of cake = $\frac{1}{2} \times \frac{3}{8}$ of cake = $\frac{3}{16}$ of cake

36.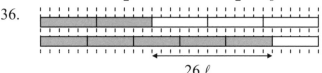

$\frac{5}{6} - \frac{2}{5} = \frac{25}{30} - \frac{12}{30} = \frac{13}{30}$ OR: Divide tank into $5 \times 6 = 30$ units

$\frac{13}{30} \longrightarrow 26 \ \ell.$ $(5 \times 5) - (2 \times 6)$ units = 13 units

$\frac{30}{30} \longrightarrow \frac{26}{13} \ \ell \times 30 = 60 \ \ell$ 13 units = 26 ℓ

The capacity of the tank is 60 ℓ.   30 units = $\frac{26}{13} \ \ell \times 30 = 60 \ \ell$

38. apples : oranges : pears   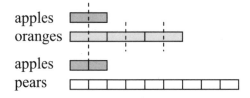

     1 : 3

 = 2 : 6

     2   : 9

     2 : 6 : 9

Total units = $2 + 6 + 9 = 17$ $\dfrac{\text{Number of pears}}{\text{Total number of fruit}} = \dfrac{9}{17}$

Copyright © 2006 SingaporeMath.com Inc., Oregon

40. Percentage of the members who are boys = 100% − 15% = 85%
    15% of the members is 18.
    15% ⟶ 18
    85% ⟶ $\frac{18}{15} \times 85 = 102$
    There are 102 boys.

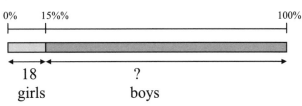

18 girls   ? boys

46. Base of first triangle = 13 cm − 5cm = 8 cm
    Height of first triangle = 2 cm
    Area of first triangle = $\frac{1}{2} \times 8 \times 2 = 8$ cm^2
    Base of second triangle = 2 cm (left side is taken as the base)
    Height of second triangle = 5 cm
    Area of second triangle = $\frac{1}{2} \times 2 \times 5 = 5$ cm^2
    Total area = 8 cm^2 + 5 cm^2 = 13 cm^2

Note: Some students may not recognize that they are given the base of the second unshaded triangle. They can, correctly, solve the problem by subtracting the two unshaded triangles in the 5 cm by 5 cm square from the area of the square. However, it is important for them to recognize they can use a side of the triangle that is not parallel to the bottom of the page as the base of the triangles. In earlier levels, students did learn that the formula for area of a triangle applies to all triangles, not just right triangles. If they are not aware of this, be sure to review area of triangles in *Primary Mathematics 5A*, unit 4.

Workbook Review 2

12. (a) $\frac{\text{Joe's weight}}{\text{Colin's weight}} = \frac{2}{3}$

    (b) 3 units = 36 kg
    1 unit = $\frac{36}{3}$ kg = 12 kg

    US edition   3d edition
    Joe   Ali
    Colin   Gopal
    36 kg

    Joe's weight = 2 × 12 kg = 24 kg
    If Joe gains 3 kg, his new weight becomes 24 + 3 = 27 kg.
    New ratio = 27 : 36 = 3 : 4 (divide both by 9)

18. 20% of $1200 = $\frac{1}{5} \times \$1200 = \$240$

    20% of $1500 = $\frac{1}{5} \times \$1500 = \$300$ OR: $1500 − $1200 = $300

    Increased savings = $300 − $240 = $60   20% of $300 = $\frac{1}{5} \times \$300 = \$60$

    He can save $60 more each month.

Copyright © 2006 SingaporeMath.com Inc., Oregon

19.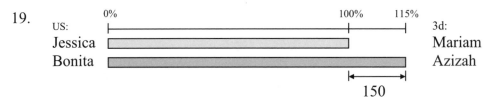

Total stamps = 215% of Jessica's stamps
15% of Jessica's stamps is 150.

$15\% \longrightarrow 150$

$215\% \longrightarrow \dfrac{150}{15} \times 215 = 2150$

They have 2150 stamps altogether.

20. Percentage of marbles that are green = $100\% - 40\% - 20\% = 40\%$
40% of the marbles is 80.

$40\% \longrightarrow 80$

$100\% \longrightarrow \dfrac{80}{40} \times 100 = 200$

There are 200 marbles.

21.

food 20%   remainder 80%

transport 30%   savings 70%

Percentage of salary saved = 70% of 80% = $\dfrac{7}{10} \times 80\% = 56\%$

23. 25 revolutions $\longrightarrow$ 5 min

1 revolution $\longrightarrow \dfrac{5}{25}$ min

30 revolutions $\longrightarrow \dfrac{5}{25} \times 30 = 6$ min

It will take 6 minutes to make 30 revolutions.

Copyright © 2006 SingaporeMath.com Inc., Oregon

Unit 5 – Speed

Objectives

- Interpret speed as distance traveled per unit of time.
- Read and write units of speed.
- Find average speed.
- Solve word problems that involve speed, distance, and time.

Suggested number of sessions: 8

	Objectives	Textbook	Workbook	Activities
Part 1 : Speed and Average Speed				**8 sessions**
53	▪ Interpret speed as the rate of distance traveled per unit of time. ▪ Read and write units of speed. ▪ Find distance when speed and time are given.	pp. 74-75, tasks 1-3 p. 81, Practice 5A, 3		5.1a
54	▪ Find speed when distance and time are given. ▪ Find average speed when distance traveled and time are given.	p. 76, tasks 4-7 p. 81, Practice 5A, 1-2	Ex. 27	5.1b
55	▪ Find time when average speed and distance are given.	p. 76, tasks 8-9 p. 81, Practice 5A, 4,10	Ex. 28	5.1c
56	▪ Solve 2-step word problems that involve speed. ▪ Draw diagrams to help solve word problems that involve speed.	pp. 77-78, tasks 10-12 p. 81, Practice 5A, 5-9	Ex. 29	5.1d
57	▪ Find average speed for different speeds traveled on different parts of a trip.	p. 79, tasks 13-14	Ex. 30	5.1e
58	▪ Solve multi-step word problems that involve speed.	p. 80, tasks 15-16	Ex. 31	5.1f
59	▪ Practice.	p. 81, Practice 5A		5.1g
60		p. 82, Practice 5B		

Copyright © 2006 SingaporeMath.com Inc., Oregon

Part 1: Speed and Average Speed	8 sessions

Objectives

- Interpret speed as the rate of distance traveled per unit of time.
- Read and write units of speed.
- Understand average speed.
- Solve 2-step word problems involving speed.
- Solve multi-step word problems involving two parts traveled at different speeds.

Material

- A chart listing distances from the school to some nearby places or towns.

Homework

- Workbook Exercise 27
- Workbook Exercise 28
- Workbook Exercise 29
- Workbook Exercise 30
- Workbook Exercise 31

Notes

Speed is a special kind of rate: the rate of distance covered per unit of time.

We write units of speed as km/h, m/min, m/s, cm/s, mi/h, etc.

Students can regard speed as a rate which involves distance and time, and solve speed problems in the same way as rate problems. Students will be deriving formulas that involve speed from rate "equations". Rate problems were studied in *Primary Mathematics 5B*. If your students have not previously done *Primary Mathematics 5B*, it would be helpful for you to go over Unit 2 from that level with the class.

In the text and workbook, students will be given two of the three quantities, speed, time, or distance, and asked to find a third.

Students will learn the formula: Distance = Speed × Time. From this they can find the speed when given the distance and time: $\text{Speed} = \dfrac{\text{Distance}}{\text{Time}}$, or the time when given the distance and speed: $\text{Time} = \dfrac{\text{Distance}}{\text{Speed}}$

Students should be able to solve these problems the same way as rate problems, rather than using one of the above three formulas. Some students may even prefer to solve them as rate problems. Allow them to use whichever approach they prefer in the workbook exercises.

Copyright © 2006 SingaporeMath.com Inc., Oregon

If the speed of a vehicle remains the same over an interval of time, then it is traveling at a uniform speed. Other terms can be used, such as steady speed, constant speed, fixed speed, or given speed. Usually, however, the speed of a car varies during a trip, so when we divide the total distance covered during the trip by the time it takes, we get the average speed for the trip.

For a trip that is divided into two or more parts, we must find the total distance traveled and the total time in order to find the whole trip's average speed. It is not possible to simply average the two speeds for each part of the trip, even if the distance is the same. This is often a point of confusion for students.

For example, a car travels for 180 km at 90 km/h and for another 180 km at 30 km/h. The average speed is not the average of 90 km/h and 30 km/h (60 km/h). For the first part of the trip (180 km at 90 km/h), the car traveled for 2 h; for the second part (180 km at 30 km/h), it traveled for 6 h. The total time for the trip is 8 h, the trip's total distance is 360 km. The average speed, then, is $\frac{360 \text{ km}}{8 \text{ h}}$ = 45 km/h (closer to 30 km/h, not right between 30 and 90, since it spent more time traveling at 30 km/h).

Students will be solving two-step and multi-step word problems which involve speed. Arrow diagrams will be used, as well as line diagrams. Line diagrams are drawn as a line between two end-points, labeled for distance, time, and speed. They are helpful since they can be labeled with all the pertinent information we begin with. Encourage students to draw diagrams similar to those shown in the text.

Speed problems are a great opportunity to bring together a variety of concepts: rates, proportion, time, decimals, fractions, problem solving skills, converting between decimals and fractions, adding and subtracting in compound units, converting between measurement units, etc. If possible, take some time to have students work through some more challenging problems from any of the supplements.

Copyright © 2006 SingaporeMath.com Inc., Oregon

Activity 5.1a   **Distance = Speed × Time**

1. Discuss the concept of speed.
    *   Students are likely to be familiar with the concept of speed. Discuss different instances where speed is used, such as the speed of a car, the speed at which a cheetah runs, the speed of an airplane, etc.
    *   Discuss reasons for setting traffic speed limits.
    *   Help students get a "feel" for relative speeds. You can compare the speed of walking to the speed in a car and how long it takes to go from, say, the school to a nearby park when walking versus driving. Or, how long it might take to drive across the country compared to flying by airplane (not including time at the airport).
    *   Remind students that a rate compares two quantities with different units to each other. Discuss some examples of rate, such as glasses of water that should be drunk *per* day, amount of rainfall *per* year, etc. Note that rates are generally given as some amount in one measurement unit *per one* of the other measurement unit. Speed is a type of rate (km/h, ft/min).
    *   If students are weak in their understanding of rate, or unfamiliar with arrow and line diagrams, you may wish to take a few days to review the concept of rate and solving rate problems. You can use the material from *Primary Mathematics 5B*, unit 4, and from the Teacher's Guide for that level.

2. Discuss the contents of **p. 74 in the textbook.**
    *   A speedometer is like a ruler wrapped into a circle or partial circle. The units are speed in terms of distance per time.
    *   Draw arrow diagrams for the information at the bottom of the page, and also express the relationship as a fraction.

        1 h ⟶ 75 km
        2 h ⟶ $75 \times 2 = 150$ km
        3 h ⟶ $75 \times 3 = 225$ km

    *   Point out that to maintain a set distance per time; that is, to keep the ratio the same, we multiply the hours by the same amount as the distance.

        $$\frac{75}{1} = \frac{150}{2} = \frac{225}{3}$$

3. Discuss **task 1, textbook p. 75.**

4. Find the distance traveled when the speed and time are given.
    *   Refer to **tasks 2-3, textbook p. 75.**
    *   Write the rate "equations", using arrows. (3rd edition has km instead of mi.)

        In 1 h, the van can travel 50 miles.
        1 h ⟶ 50 mi
        2 h ⟶ $50 \times 2 = 100$ mi
        It can travel 100 miles in 2 hours.
        Speed (**50** mi/h) × Time (**2** h) = Distance (**100** mi)

    *   Remind students that speed is the distance traveled per **one** unit of time.
    *   Point out that when we are given the speed, we are told the relationship between a specific distance and one unit of time.

- From this, we can find the distance traveled for any time period using the distance for 1 unit of time. We multiply the distance by the same number that our 1 unit of time is multiplied by.

1 min $\longrightarrow$ 40 m
5 min $\longrightarrow$ 40 × 5 = 200 m
He can swim 200 m in 5 min.
Speed (**40** m/min) x Time (**5** min) = Distance (**200** m)

- So distance for a new time is the speed for one unit of time × the new time, or:
Distance = Speed × Time.

Distance = Speed × Time

5. Have students do **problem 3, Practice 5A, textbook p. 81.**

6. Have students do some problems where they need to be aware of the unit of time and convert it to the same unit used in the speed:
   - A car is traveling at a speed of 80 km/h.
     How far can it travel in 4 hours? (320 km)
     How far can it travel in $2\frac{1}{2}$ hours? (200 km)
     How far can it travel in 90 minutes? (120 km)
     How far can it travel in 0.5 hours? (40 km)
     How far can it travel in 15 minutes? (20 km)
   - Similarly, ask a student to estimate his or her speed on a bicycle. Then ask how far he or she can go in a certain number of hours (include mixed fractions or decimals for number of hours) or in a certain number of minutes.

Activity 5.1b   **Speed = Distance ÷ Time**

1. Find the speed when the distance and time are given.
   - Refer to **tasks 4-5, textbook p. 76**.
   - For task 4, tell students we are given the relationship between two quantities, meters (distance) and seconds (time), and need to find the meters that the bullet travels for a new time (1 second).
   - Write the rate "equation" with arrows for 2 s and 420 m. Since we will want to find a new distance, we put the distance to the right of the arrow.

In 2 seconds, the bullet travels 420 m.
2 s $\longrightarrow$ 420 m
1 s $\longrightarrow$ $\frac{420}{2}$ = **210** m
$\frac{\text{Distance (}\mathbf{420}\text{ m)}}{\text{Time (}\mathbf{2}\text{ s)}}$ = Speed (**210** m/s)

   - To find the distance for 1 s, we divide both sides of the "equation" by 2. Write the new rate "equation".
   - In 1 second, the bullet travels 210 m. This gives us the bullet's speed, 210 m/s.

Copyright © 2006 SingaporeMath.com Inc., Oregon

- So, dividing the distance traveled by the time, we get the speed, which is distance per unit time.

- For task 5, ask students to tell you the relationship given (distance in meters to time in seconds) and what we need to find (distance for 1 second will give us the speed).

- Ask students to write the rate diagram.

- To get speed, we divide distance by time. $\text{Speed} = \dfrac{\text{Distance}}{\text{Time}}$

In 30 seconds, he runs 150 m.

$30 \text{ s} \longrightarrow 150 \text{ m}$

$1 \text{ s} \longrightarrow \dfrac{150}{30} = \mathbf{5} \text{ m}$

$\dfrac{\text{Distance } (\mathbf{150} \text{ m})}{\text{Time } (\mathbf{30} \text{ s})} = \text{Speed } (\mathbf{5} \text{ m/s})$

$\text{Speed} = \dfrac{\text{Distance}}{\text{Time}}$

2. Have students do **problems 1-2, Practice 5A, textbook p. 81.**

3. Discuss average speed.
   - Tell students that when we drive a car, even on a highway, we do not always go at the same speed. We might speed up to pass, or slow down when traffic piles up. But if we go a total distance of 60 miles (or kilometers) in one hour, we can say that the *average* speed was 60 mi/h (or km/h), even though we might have been going 70 mi/h for part the trip, and 45 mi/h for another part of it.
   - We can think of average speed as the overall summary of a trip during which we went faster some of the time and slower some of the time. We know how far we went in all, and how long that took. By dividing this total distance by the total time, we find the average speed.

   $\text{Average Speed} = \dfrac{\text{Total Distance}}{\text{Total Time}}$

   - Display a chart with distances from the school to various locations. Have students estimate the time it would take to get to those places and then find the average speed for the trip.

4. Have students do **tasks 6-7, textbook p. 76**

Workbook Exercise 27

| **Activity 5.1c** | **Time = Distance ÷ Speed** |

1. Find the time when given the distance and speed.
   - Tell students we have found distance when given the speed and time, and speed when given the distance and time. Have students tell you those formulas and write them on the board.
   - Tell them that now we will find time when the distance and speed are given.

$\text{Distance} = \text{Speed} \times \text{Time}$

$\text{Speed} = \dfrac{\text{Distance}}{\text{Time}}$

Copyright © 2006 SingaporeMath.com Inc., Oregon

- Write the problem shown at the right on the board. We need to find the time it takes to go 240 km at a speed of 60 km/h.

> A train traveled 240 km at an average speed of 60 km/h. How long did the trip take?

- Guide students in solving this as a rate problem using the unitary method. We know the distance traveled in 1 h. We are given a new distance, 240 km, and need to find the time this takes.

$$60 \text{ km} \longrightarrow 1 \text{ h}$$

$$1 \text{ km} \longrightarrow \frac{1}{60} \text{ h}$$

- The expression, $\frac{1}{60} \times 240$, can be rewritten as $\frac{240}{60}$. This is the distance divided by the speed.

$$240 \text{ km} \longrightarrow \frac{1}{60} \times 240 \text{ h} = 4 \text{ h}$$

$$\frac{240}{60} = 4 \text{ h}$$

- So we can find the new time by dividing the new distance by the speed, which is distance per 1 unit of time.

$$\frac{\text{Distance } (\mathbf{240} \text{ km})}{\text{Speed } (\mathbf{60} \text{ km/h})} = \text{Time } (\mathbf{4} \text{ h})$$

$$\text{Time} = \frac{\text{Distance}}{\text{Speed}}$$

$$\text{Time} = \frac{\text{Distance}}{\text{Speed}}$$

2. Discuss **tasks 8-9, textbook p. 77**.

- In task 8, point out that the expression, $\frac{245}{70}$ hours, is $\frac{\text{Distance}}{\text{Speed}}$.

- You may want to have students draw the rate diagram using arrows for task 9.

3. Review the three formulas for speed.
   - Write the three equations on the board.
   - Tell students that if they remember Distance = Speed × Time, they can use that to find the other two formulas. It is also easy to remember $\text{Speed} = \frac{\text{Distance}}{\text{Time}}$ since speed is defined as distance per unit time.

$$\text{Distance} = \text{Speed} \times \text{Time}$$

$$\text{Speed} = \frac{\text{Distance}}{\text{Time}}$$

$$\text{Time} = \frac{\text{Distance}}{\text{Speed}}$$

   - Tell students that they can always solve problems that involve speed as rate problems.

4. Have students do **problems 4 and 10, Practice 5A, textbook p. 81**.

   Workbook Exercise 28

Activity 5.1d   **Speed diagrams**

1. Introduce speed diagrams.
   - Refer to **task 10(a), textbook p. 77**. (The 3rd edition has km instead of mi.)
     - Tell students that this is a fairly easy problem. As problems become more difficult, however, it will be helpful to have a way to diagram the pertinent information. Remind them that we have used bar diagrams to diagram word problems involving whole numbers and fractions, and line diagrams to diagram rate problems. We can also use line diagrams in solving problems that involve speed. We draw a line to show the path from

Copyright © 2006 SingaporeMath.com Inc., Oregon

one place to the next, and then label it with information from the problem. We then use the diagram to help us solve the problem.

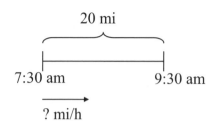

o Draw a line and put dots or bars at both ends. These are the starting point and the ending point of the bicycle trip.
o Guide students in labeling the drawing with information from the problem. We can indicate the total distance (20 mi), the start time (7:30 am), the end time (9:30 am), and the quantity we want to find (speed).
o Ask students what two things we need to know to find the speed. We need the distance (20 mi) and the time (2 h) it takes to go that distance. Let them tell you that the average speed is $\frac{20}{2}$ = 10 mi/h.

- Refer to **task 10(b), textbook p. 77**.

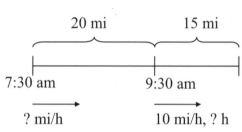

o Guide students in diagramming the information given. We can add to the previous diagram that we drew for 10(a).
o Now, we need to find the new end time. To find time, we need to know the other two quantities, speed and distance. Using the diagram, we can easily see what we need to find, and what we already know.
o Have students solve the problem.

2. Discuss **tasks 11-12, textbook p. 78**.
- Ask students to tell you what we need to find, and what we need to know to find it. Relate the information in the diagrams to the information in the word problem. Use the diagram to determine an approach for solving the problem.
- In task 11, we need to find the time for driving back. To find time, we need to know speed and distance. We are given the speed (60 km/h), but not the distance. However, we can find the distance from other information in the problem. We know the speed (75 km/h) and time (4 h) for the trip from P to Q, and so can find the distance (75 × 4 = 300 km), which is the same as the distance for the return trip. Knowing the distance, and the speed for the return trip (60 km/h) we can now find the time for the return trip (300 km ÷ 60 km/h = 5 h).
- In task 12, we need to find the average speed for the van. To find speed, we need to know the distance and the time for the van. The time is given (5 h). We can find the distance from the motorcyclist's speed (35 km/h) and time (7 h). The distance is 35 × 7 = 245 km. We now have a distance and a time for the van, and can find its average speed (245 km ÷ 5 h = 49 km/h).
- Point out that by having all the information in a diagram, we can easily see what we know, what we need to find, and what we can use to find it.

Copyright © 2006 SingaporeMath.com Inc., Oregon

3. Have students do **problems 5-9, Practice 5A, textbook p. 81,** as time permits.
   - Ask them to draw diagrams.
   - Get some to show their diagrams and explain their solutions.

Workbook Exercise 29

Activity 5.1e   **Word problems**

1. Discuss **task 13, textbook p. 79**.
   - Have students read the problem and draw a diagram, or relate the information in the given diagram to the information in the problem. Point out that the line is divided into two parts, one for each speed.
   - Tell students that we cannot simply average the two speeds to find the average speed. Only two hours were spent going at 80 km/h, but three hours were spent going 70 km/h (even those two speeds are averages for the hour). Ask students whether the average speed would be closer to 70 km/h or 80 km/h. It will be closer to 70 km/h, since he went at that speed for the greater part of the trip. You can have them find the average by averaging the speed for each hour:
     $$\frac{80+80+70+70+70}{5} = \frac{370}{5} = 74$$
   - To find average speed we need to divide the *total* distance by the *total* time. Ask students how we can find the total distance. Since we are given the time and speed for each part, we can find the distance for each part, and add them. Ask students to help you find the total time and then the average speed.
     Total distance traveled = (80 km/h × 2 h) + (70 km/h × 3 h) = 370 km
     Total time taken = 2 h + 3 h = 5 h
     Average speed = $\dfrac{370 \text{ km}}{5 \text{ h}}$ = 74 km/h

2. Discuss **task 14, textbook p. 79**.
   - This problem is similar to the previous one, except that for each part of the trip we are given the distance and the speed, rather than the time and the speed. So we need to calculate the time for each part of the trip in order to find the total time.
     Total distance traveled = 36 km + 96 km = 132 km
     Total time taken = $\dfrac{36 \text{ km}}{54 \text{ km/h}} + \dfrac{96 \text{ km}}{72 \text{ km/h}} = \dfrac{2}{3}$ h $+ \dfrac{4}{3}$ h = 2 h
     Average speed = $\dfrac{132 \text{ km}}{2 \text{ h}}$ = 66 km/h

3. Provide some additional problems for practice, such as the following:
   ➢ Rosalind took 36 minutes to drive from Town A to Town B at an average speed of 80 km/h. She then took another 24 minutes to drive from Town B to Town C at an average speed of 100 km/h. What was her average speed for the whole trip? (88 km/h)
     ○ If necessary, point out that, since the speed is measured in km/h, they will first need to change the time from minutes to hours.

Copyright © 2006 SingaporeMath.com Inc., Oregon

➢ Pierre took 5 hours to travel from Town A to Town B. He traveled at an average speed of 90 km/h for the first 270 km. He then increased his speed by 20 km/h for the remaining part of the journey. What was his average speed for the whole journey? (98 km/h)
  o If students need a hint, tell them they have to first find the time the first part of the trip takes. Then they can find the time for the second part of the trip by subtracting that from the total time. They will then have a speed and a time for the second part of the trip, and can find the total distance.

Workbook Exercise 30

Activity 5.1f   **Word problems**

1. Discuss **task 15, textbook p. 80**.
   • To find the average speed, we need to find the total distance and the total time.
   • How do we find the total distance?
     o We know the distance for $\frac{1}{3}$ of the trip. From this we can find the total distance.

     $\frac{1}{3}$ of the distance = 120 km

     $\frac{3}{3}$ of the distance = $120 \times 3 = 360$ km

   • How do we find the total time?
     o We are given the time for both parts, so we can add to find the total time.

     Total time taken = 3 h + 2 h = 5 h

   • Now we can find the average speed.

     Average speed = $\frac{360 \text{ km}}{5 \text{ h}} = 72$ km/h

2. Discuss **task 16, textbook p. 80**.
   • Have students draw a diagram or copy the one in the text.
   • To find the average speed for the remaining trip, we need to find the remaining distance and time.
   • In these problem, if we don't immediately see a solution, we can start by seeing how many of the three quantities we can find for each part of the trip and for the whole trip.
   • We have the time and speed for the whole trip, so we can find the total distance (240 km).
   • We can also find the distance for each part, since we know what fraction each part is of the whole. Add the distance for each part of the trip to the diagram.

     $\frac{1}{5}$ of the trip = $\frac{1}{5} \times 240$ km = 48 km

     $\frac{4}{5}$ of the trip = $\frac{4}{5} \times 240$ km = 192 km

   • We can also find the time for the first part of the trip, since we now have both the speed and the distance. Add that time to the diagram.

     Time taken for the first part = $\frac{192 \text{ km}}{64 \text{ km/h}} = 3$ h

   • Now we see that we can find the time for the last part of the trip, since we have the total time and the time for the first part.

     Time taken for the remaining part = 4 h – 3 h = 1 h

   • Now we have both time and distance for the remaining trip, and can find the average speed.

     Average speed for remaining part = $\frac{48 \text{ km}}{1 \text{ h}} = 48$ km/h

Copyright © 2006 SingaporeMath.com Inc., Oregon

3. Provide some additional problems for practice, such as the following ones, and then have students discuss their solution. They should draw diagrams and, if they are at a loss for how to solve the problem, fill in as much information as possible until they see a solution. (For 3rd edition, change miles to kilometers.)

➤ Jack left Town A at 2:00 pm for Town B. He drove at an average speed of 51 mi/h for the first 3 hours and 20 minutes. Then he drove another 4 hours, at an average speed of 45 mi/h, to reach Town B. Nicole left Town A $1\frac{1}{2}$ h earlier than Jack left Town A. At 5:30 pm, Nicole still had $\frac{3}{7}$ of the distance to go to Town B. When did Nicole reach Town B, assuming she traveled the whole distance at the same average speed?

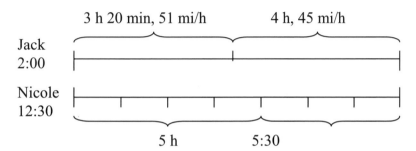

$3 \text{ h } 20 \text{ min} = 3\frac{1}{3} \text{ h} = \frac{10}{3} \text{ h}$

Total distance $= (\frac{10}{3} \text{ h} \times 51 \text{ mi/h}) + (4 \text{ h} \times 45 \text{ mi/h}) = 170 \text{ mi} + 180 \text{ mi} = 350 \text{ mi}$

Nicole went $\frac{4}{7}$ of the way in 5 h.

Distance for those 5 h $= \frac{4}{7} \times 350 = 200 \text{ mi}$

Speed $= \frac{200}{5} = 40 \text{ mi/h}$

Remaining distance $= 350 - 200 = 150 \text{ mi}$

Remaining time $= \frac{150}{40} = 3\frac{3}{4} \text{ h} = 3 \text{ h } 45 \text{ min}$

Nicole reached Town B at 9:15 pm

➤ Sean took 3 hours and 45 minutes to complete 60% of his trip at an average speed of 60 mi/h. His average speed for the whole trip was 45 mi/h. What was his time for the last 40% of the trip? (4 h 35 min)

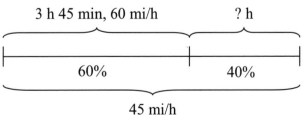

Copyright © 2006 SingaporeMath.com Inc., Oregon

To find the speed of the second part of his trip, we first need to find the distance and time for the last 40% of the trip.

$$3 \text{ h } 45 \text{ min} = 3\frac{3}{4} \text{ h} = \frac{15}{4} \text{ h}$$

Distance for first 60% = $\frac{15}{4}$ h $\times$ 60 mi/h = 225 mi

Distance for last 40%:

60% of total = 225 mi	or	$\frac{3}{5}$ of total = 225 mi
1% of total = $\frac{225}{60}$ mi		$\frac{1}{5}$ of total = $\frac{225}{3}$ mi
40% of total = $\frac{225}{60} \times 40 = 150$ mi		$\frac{2}{5}$ of total = $\frac{225}{3} \times 2 = 150$ mi

Total distance = 225 mi + 150 mi = 375 mi

Total time = $\frac{375}{45} = \frac{75}{9} = 8\frac{1}{3}$ h

Time for last part of the trip = $8\frac{1}{3}$ h $- 3\frac{3}{4}$ h $= 4\frac{7}{12} = 4$ h 35 min

Workbook Exercise 31

Activity 5.1g   **Practice**

1. Have students do problems from **Practice 5A, textbook p. 81 and Practice 5B, textbook p. 82.** Discuss their solutions.

    Possible solutions for problems 5-7 in Practice 5B:

    #5. Distance for second part
         = 12 km – 3 km = 9 km

         Total time = $\frac{3 \text{ km}}{4 \text{ km/h}} + \frac{9 \text{ km}}{6 \text{ km/h}}$

         $= 2\frac{1}{4}$ h = 2 h 15 min

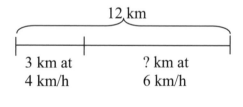

    #6. (a) $\frac{1}{5}$ of the distance = 30 km

              $\frac{5}{5}$ of the distance = $30 \times 5 = 150$ km

         (b) Total time = 2 h + 1 h = 3 h

              Average speed = $\frac{150 \text{ km}}{3 \text{ h}} = 50$ km/h

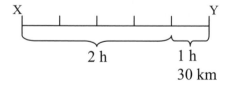

Copyright © 2006 SingaporeMath.com Inc., Oregon

#7.   (a) $\frac{2}{3}$ of the distance = 240 km

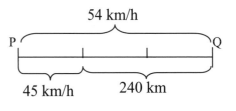

$\frac{1}{3}$ of the distance = $\frac{240}{2}$ = 120 km

$\frac{3}{3}$ of the distance = 120 × 3 = 360 km

(b)   Use the total distance and average speed to find the total time.

Total time = $\frac{360 \text{ km}}{54 \text{ km/h}}$ = $\frac{20}{3}$ h

We have both speed (45 km/h) and distance (120 km) for the first part of the trip. Find the time and distance for the second part of the trip.

Time for first part = $\frac{120 \text{ km}}{45 \text{ km/h}}$ = $\frac{8}{3}$ h

Time for second part = $\frac{20}{3}$ h $-$ $\frac{8}{3}$ h = $\frac{12}{3}$ = 4 h

Average speed for second part = $\frac{240 \text{ km}}{4 \text{ h}}$ = 60 km/h

Review

Objectives

• Review previous material.

Suggested number of sessions: 10

	Objectives	Textbook	Workbook	Activities
Review				**10 sessions**
61-70	▪ Review	pp. 83-86, Review D pp. 87-92, Review E US pp. 93-96, Review F	Review 3 Review 4	

Possible solutions to selected problems:

Review D (pp. 83-86)

28. 2 units = 20

$8 \text{ units} = \frac{20}{2} \times 8 = 80$

There are 80 children.

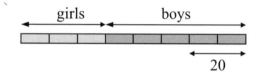

29. $\frac{1}{4}$ of the book = 1 unit

2 units = 36 + 6 = 42

$1 \text{ unit} = \frac{42}{2} = 21$

She read 21 pages on Sunday.

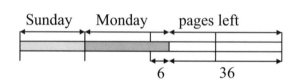

32. *Before*: All their money = 3 units = $240

$1 \text{ unit} = \$\frac{240}{3} = \$80$

Juan's money = 2 × $80 = $160

Tom's money = $80

After: Juan's money = $160 - $20 = $140

Tom's money = $80 + $20 = $100

New ratio = 140 : 100 = 7 : 5

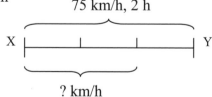

37. (a) Distance = 75 km/h × 2 h = 150 km (3rd edition: Pablo is Hamid)

Distance Pablo traveled in 2 h = $\frac{2}{3} \times 150 = 100$ km

Pablo's speed = $\frac{100 \text{ km}}{2 \text{ h}} = 50$ km/h

(b) Distance for last third = 50 km

Speed for last third = 50 + 10 = 60 km/h

Time = $\frac{50 \text{ km}}{60 \text{ km/h}} = \frac{5}{6}$ h = $\frac{5}{6} \times 60$ min = 50 min

Copyright © 2006 SingaporeMath.com Inc., Oregon

Review E, pp. 87-92

27.

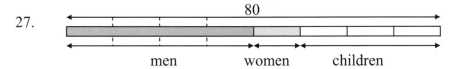

men   women   children

There are 5 units of adults and
3 units of children. The difference is 2
units.

8 units = 80
1 unit = 10
2 units = 2 × 10 = 20

There are 20 more adults than children.

or:

Number of men = $\frac{1}{2} \times 80 = 40$

Remainder = $80 - 40 = 40$

Number of women = $\frac{1}{4} \times 40 = 10$

Number of children = $40 - 10 = 30$

Total adults = $40 + 10 = 50$

Adults $-$ children = $50 - 30 = 20$

29. bottle + 4 units of water = 2.4 kg
     bottle + 1 unit of water = 1.2 kg
     3 units of water = $2.4 - 1.2 = 1.2$ kg
     1 unit of water = $\frac{1.2}{3} = 0.4$ kg

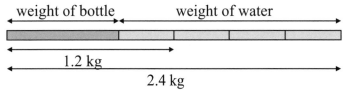

weight of bottle = weight of bottle with 1 unit of water − weight of 1 unit of water
                 = $1.2 - 0.4 = 0.8$ kg

31. Number of children stays the same.
     Before: adults : children = 8 : **3**
     After: adults : children = 2 : 1 = 6 : **3**
     2 units of adults leave.
     2 units = 10
     3 units = $\frac{10}{2} \times 3 = 15$
     There were 15 children on the ferry.

Before:
Adults
Children
After:
Adults   10
Children

32. (a) John : David
           3 : **4**
               David : Paul
                  2 : 1
               = **4** : 2
         John : David : Paul
            3 : 4 : 2

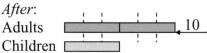

John
David
Paul
                    $60

     (b) 2 units = $60
          3 units = $\frac{60}{2} \times 3 = \$90$
          John has $90.

Copyright © 2006 SingaporeMath.com Inc., Oregon

37. 54 is 30% of the girls. We want to find 230% of the girls, which is all the children.

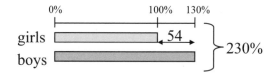

$30\% \longrightarrow 54$

$1\% \longrightarrow \dfrac{54}{30}$

$230\% \longrightarrow \dfrac{54}{30} \times 230 = 414$ children

Or:

$30\% \longrightarrow 54$

$10\% \longrightarrow \dfrac{54}{3} = 18$

$100\% \longrightarrow 180$ girls

$230\% \longrightarrow 180 + 180 + 54 = 414$ children

There are 414 children.

39. Distance for first part of the trip = 75 km/h × 2 h = 150 km

Total distance = 150 × 2 = 300 km

Total time = $\dfrac{300 \text{ km}}{60 \text{ km/h}} = 5$ h

Time for last part = 5 h − 2 h = 3 h

Speed for last part = $\dfrac{150 \text{ km}}{3 \text{ h}} = 50$ km/h

40. Area of Δ1 = $\dfrac{1}{2} \times 18 \times 11 = 99$ cm^2

Area of Δ2 = $\dfrac{1}{2} \times 18 \times 5 = 45$ cm^2

Area of Δ3 = area of Δ1 − area of Δ2
                   = 99 − 45 = 54 cm^2

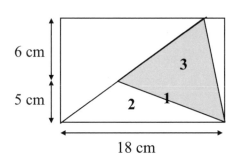

^{US}Review F, pp. 93-96

5.   (a) 5 yd 1 ft ÷ 2 = (4 yd + 3 ft + 1 ft) ÷ 2 = (4 yd ÷ 2) + (4 ft ÷ 2) = 2 yd 2 ft
     (b) 13 lb 8 oz ÷ 6 = (12 lb + 16 oz + 8 oz) ÷ 6 = (12 lb ÷ 6) + (24 oz ÷ 6) = 2 lb 4 oz
     (c) 19 gal 2 qt ÷ 3 = (18 gal + 4 qt + 2 qt) ÷ 3 = (18 gal ÷ 3) + (6 qt ÷ 3) = 6 gal 2 qt
     (d) 5 ft 10 in. ÷ 7 = (60 in. + 10 in.) ÷ 7 = 70 in. ÷ 7 = 10 in.

8.   (a) $4\dfrac{1}{4}$ lb = (4 × 16 oz) + ($\dfrac{1}{4}$ × 16 oz) = 68 oz

     (b) $2\dfrac{2}{3}$ yd = (2 × 36 in.) + ($\dfrac{2}{3}$ × 36 in.) = 72 in. + 24 in. = 96 in.

     (c) $5\dfrac{1}{2}$ gal = 20 qt + 2 qt = 22 qt = 44 pt = 88 c

Copyright © 2006 SingaporeMath.com Inc., Oregon

23. $\frac{5}{7}$ of the tank can hold 3.5 gal

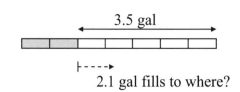
3.5 gal

$\frac{1}{7}$ of the tank holds 3.5 gal ÷ 5 = 0.7 gal

2.1 gal fills to where?

$2.1 ÷ 0.7 = 3$

2.1 gal is $\frac{3}{7}$ of the tank.

The tank was $\frac{2}{7}$ filled; adding 2.1 more gallons makes it $\frac{5}{7}$ filled.

Workbook Review 3

15.
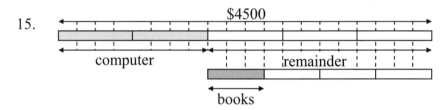
$4500

computer   remainder

books

$\frac{3}{5} \times \$4500 = \$2700$ (the remainder)

or: $\frac{1}{4}$ of the remainder $= \frac{1}{4} \times \frac{3}{5} = \frac{3}{20}$

$\frac{1}{4} \times \$2700 = \$675$ (cost of books)

$\frac{3}{20}$ of $4500 = $\frac{3}{20} \times \$4500 = \$675$

OR: Divide total into 20 units.

20 units = $4500

1 unit = $\frac{4500}{20}$

3 units = $\$\frac{4500}{20} \times 3 = \$675$

He spent $675 on books.

16.
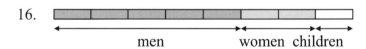
men   women children

Divide the total into 8 units.   or: Fraction of participants that are children

Men = 5 units

Women = 2 units $= 1 - \frac{5}{8} - \frac{1}{4} = \frac{8}{8} - \frac{5}{8} - \frac{2}{8} = \frac{1}{8}$

Children = 1 unit

$\frac{1}{8} \times 100\% = 12.5\%$ Percentage $= \frac{1}{8} \times 100\% = 12.5\%$

12.5% of the participants are children.

Copyright © 2006 SingaporeMath.com Inc., Oregon

18.

3 h, 80 km/h

A ⊢————→———————————⌐ B
                    ←——
                60 km/h
                              1:00 pm

Distance = 80 km/h × 3 h = 240 km

Time for return = $\dfrac{240 \text{ km}}{60 \text{ km/h}}$ = 4 h

4 h after 1:00 p.m. is 5:00 p.m.
He arrived back at Town A at 5:00 p.m.

23.   The number of girls stays the same.

        boys : girls

Before   3 : 2 = 9 : **6**

After   2 : 3 = 4 : **6**

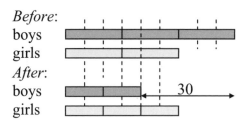

Before:
boys
girls
After:
boys   30
girls

Number of units of boys that leave = 9 − 4 = 5

5 units = 30 boys

9 units = $\dfrac{30}{5}$ × 9 = 54 boys

There were 54 boys at first.

24.

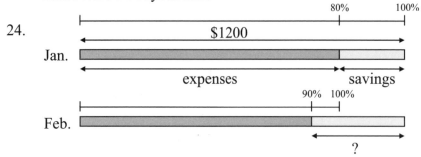

                                 80%   100%

            $1200

Jan.

         expenses   savings

                         90% 100%

Feb.

                         ?

Expenses in Jan. = 80% of $1200 = $\dfrac{80}{100}$ × $1200 = $960

Expenses in Feb. = 90% of $960 = $\dfrac{90}{100}$ × $960 = $864

Savings in Feb. = $1200 − $864 = $336

25.   8:30 a.m.   1:30 p.m.

P ⊢————→ Peter 60 km/h   ⌐ Q
  ⊢——————→

   9:30 a.m.   John ? km/h   1:30 p.m.

Peter's time = 8:30 am to 1:30 pm = 5 h

John's time = 9:30 am to 1:30 pm = 4 h

Distance = Peter's speed × Peter's time = 60 km/h × 5 h = 300 km

John's speed = $\dfrac{300 \text{ km}}{4 \text{ h}}$ = 75 km/h

Copyright © 2006 SingaporeMath.com Inc., Oregon

Workbook Review 4

7. $\frac{2}{3}$ of a number = 0.3

     $\frac{3}{3}$ of a number = $\frac{0.3}{2} \times 3 = 0.45$

15.   (3ʳᵈ edition: Adam is Aziz)
      Peter's stamps stay the same.
           Adam : Peter
      Before   5 : 6 = 5 : **6**
      After   1 : 2 = 3 : **6**

      Adam gave away 40 stamps, which
      Is 2 units
      2 units = 40

      5 units = $\frac{40}{2} \times 5 = 100$

      Adam had 100 stamps at first.

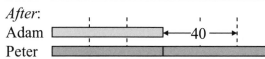

22.
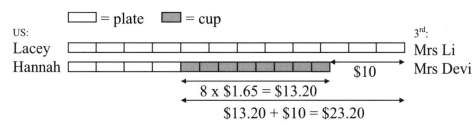

      Cost of cups = $8 \times \$1.65 = \$13.20$
      Cost of 8 plates = $\$13.20 + \$10 = \$23.20$

      Cost of 1 plate = $\$\frac{23.20}{8} = \$2.90$

23.   Amount David and Ben received
      $= 2 \times \$36 = \$72$
      Amount Ben received
      $= \frac{1}{3} \times \$72 = \$24$

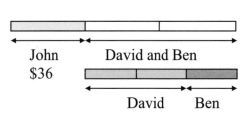

Copyright © 2006 SingaporeMath.com Inc., Oregon

24.   Laura's share = $16
      Reagan's share = $4 more than Laura's share = $4 + $16 = $20
      Laura's and Reagan's share was $16 + $20 = $36
      Since Bridget's share was 40%, Laura's and Reagan's share was 100% − 40% = 60%
      So, 60% of the money is $36.

      $100\% \rightarrow \$\dfrac{36}{60} \times 100 = \$60$

      They shared $60.

25.   Distance for first $\dfrac{3}{5}$ of the journey

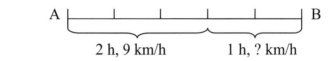

      $= 9 \text{ km/h} \times 2 \text{ h} = 18 \text{ km}$

      Distance for $\dfrac{1}{5}$ of the journey $= \dfrac{18}{3}$ km

      Distance for last $\dfrac{2}{5}$ of the journey $= \dfrac{18}{3} \times 2 = 12$ km

      Average speed for last part $= \dfrac{12 \text{ km}}{1 \text{ h}} = 12$ km/h

Copyright © 2006 SingaporeMath.com Inc., Oregon

Textbook Answer Key

Unit 1 - Algebra

Part 1: Algebraic Expressions (pp. 6-13)

1. (a) 13 years   (b) (x + 8) years
2. (a) $8   (b) $(m - 2)
3. (b) 3
4. (b) 32   (c) 44
5. (b) 21
7. (a) 12
8. (a) 4
9. (a) 10   (b) 16   (c) 9
    (d) 0   (e) 24   (f) 60
    (g) 3   (h) 1   (i) $\frac{1}{2}$
10. (b) 53
11. 7
12. (b) 19
13. 4
14. (a) 7   (b) 10
15. (a) 9   (b) 18   (c) $\frac{9}{5}$
16. (a) 8   (b) 9   (c) 12
17. (a) $6\frac{2}{3}$ (b) 23   (c) 7
    (d) $3\frac{2}{3}$ (e) $5\frac{1}{2}$ (f) 3
    (g) 47   (h) 130   (i) 120
18. (b) x
19. (b) 6r   (c) 3r + 3 (d) 6r + 3
21. (a) 9a   (b) 3c   (c) 6k
    (d) 2x + 6 (e) 5m + 7 (f) 7s + 10
    (g) 5y + 3 (h) m + 1   (i) 6r
    (j) 4p + 2 (k) 6w + 13 (l) 3h

Practice 1A (p. 14)

1. (a) 17   (b) 29   (c) 14
    (d) 12   (e) 2   (f) $\frac{1}{4}$
    (g) $\frac{3}{4}$ (h) 20   (i) 32
    (j) 44   (k) 6   (l) 2
2. (a) 3x   (b) 7x   (c) 2p
3. (a) 3p   (b) 5r   (c) f

4. (a) c   (b) 4k + 7   (c) 7n + 5
5. (a) 5g + 2   (b) 6x + 3   (c) 10
6. (a) $(y + 1)   (b) $9
7. (a) 3x m   (b) 27 m
8. (a) 3x + 4   (b) 16
9. (a) $\frac{50 - y}{2}$ (b) 6

Unit 2 - Solid Figures

Part 1: Drawing Solid Figures (pp. 15-16)

1. C
2. A. 5 faces, 2 triangles
    B. 4 faces, 4 triangles
    C. 5 faces, 4 triangles
3. D

Part 2: Nets (pp. 17-20)

2. A, D   3. B

Unit 3 - Ratio

Part 1: Ratio and Fraction (pp. 21 - 28)

    2 : 3
1. (b) 5 : 3   (d) 4 : 3 : 5
2. (b) 2 : 3 : 4
3. (a) 7 : 9   (b) 7 : 9 : 16
4. 1 : 3 : 2
5. 3 : 2 : 10
6. (d) $\frac{5}{7}$ (f) $\frac{5}{2}$
7. (a) $\frac{4}{5}$ (b) $\frac{5}{4}$
8. $\frac{5}{3}$
9. (b) 5 : 6   (c) $\frac{6}{5}$ (d) $\frac{5}{6}$
10. (a) $\frac{5}{3}$ (b) $\frac{3}{10}$
11. (a) 3 : 1   (b) 1 : 3   (c) $\frac{1}{3}$

Copyright © 2006 SingaporeMath.com Inc., Oregon

12. (a) 3 : 4   (b) 4 : 3   (c) $\dfrac{4}{3}$

13. 450

14. 45

15. 8 : 4 : 5

16. $\dfrac{2}{5}$

Practice 3A (p. 29)

1. (a) 10 : 3   (b) $\dfrac{10}{3}$

2. (a) $\dfrac{3}{4}$ (b) $\dfrac{4}{3}$

3. (a) $\dfrac{2}{3}$ (b) $\dfrac{1}{3}$

4. 2 : 3

5. 3 : 2

6 (a) 3 : 4   (b) $\dfrac{1}{4}$ (c) 36

7. 104

8. (a) 9 : 3 : 5 (b) 68

Part 2: Ratio and Proportion (pp. 30-32)

50, 18

1. (a) 8, 32   (b) 1 : 2

2. (a) 1 : 3   (b) 18 ℓ   (c) 5 ℓ

3. (a) 8 ℓ   (b) 6 ℓ

Practice 3B (p. 33)

1. (a) 2 : 5 (b) 200 g

2. (a) 4 : 3 (b) 56

3. (a) 15 (b) 4

4. ^{US} 8 gal ^{3d} 8 ℓ

5. 42

6. $24

7. $40

8. 1400

9. 16 cm

Part 3: Changing Ratios (pp. 34-37)

Peter had 32 stamps, he now has 40 stamps.

1. 1 : 2

2. Before: 60. New ratio: 4 : 5

3. 2 : 5

4. $60

5. $240

6. $39

Practice 3C (p. 38)

1. 6 : 5

2. $30

3. $90

4. $51

5. 3 : 11

6. (a) 8   (b) 16 : 13

7. (a) 30   (b) 1 : 4

8. (a) $80   (b) $200

Review A (pp 39-42)

1. (a) 40,580   (b) 2,070,000

2. (a) six hundred thousand, two hundred thirty
   (b) eight million, five thousand

3. 109,000

4. $214,000

5. (a) 45   (b) 1000

6. (a) 16   (b) 38
   (c) 9   (d) 6

7. 1, 2, or 4

8. 24

9. (a) 0.875   (b) 4.67

10. (a) $\dfrac{3}{500}$ (b) $1\dfrac{4}{5}$

11. (a) 9   (b) 3
    (c) 10; 4   (d) 2

12. 75

13. (a) 32   (b) 26   (c) 42
    (d) 16   (e) 25   (f) 72

14. (a) $10y$ (b) $7y + 7$

15. 56

16. 54.6 kg

17. $2.70

18. $4.50

19. 1 kg

20. $\dfrac{5}{16}$ ℓ

21. 80

22. 5 ℓ

23. (a) $\dfrac{1}{2}$ (b) $500

24. 96

Copyright © 2006 SingaporeMath.com Inc., Oregon

25. $\dfrac{5}{4}$

26. \$10

27. (a) 2 : 3 : 8 (b) 65

28. \$240

29. Aziz: \$30; Osman: \$15

29. (a) 800   (b) 25 min

31. \$2.70

32. \$12.60

33. 5

34. A and D

35. 44 cm

36. 112 cm^2

Review B (pp. 43-46)

1. (a) $\dfrac{5}{8}$ (b) $\dfrac{2}{5}$

2. $1\dfrac{1}{5}$

3. (a) 10   (b) 2

4. (a) 500   (b) 1000
    (c) 0.49   (d) 0.024

5. 6

6. (a) 36   (b) $\dfrac{25}{36}$

7. 0.83

8. $\dfrac{3}{5}$

9. (a) 1400 ml   (b) 125 cm

10. (a) 1720 m   (b) 2 h 5 min

11. 4:10 p.m.

12. (a) 328,000   (b) 5,000
     (c) 240,000   (d) 160

13. 1.7 ℓ

14. 47.9 kg

15. (a) (17 + m) years
     (b) 37 years

16. (a) (20 + 2x) cm
     (b) 32 cm

17. $\dfrac{16}{19}$

18. \$72.50

19. \$1.50

20. 19.5 kg

21. 400 g

22. 120

23. \$0.90

24. 120

25. (a) 262.50   (b) \$237.50

26. 6

27. 2 : 4 : 3

28. ^{US} 54 in. $^{\text{3d}}$ 54 cm

29. (a) $\dfrac{5}{11}$ (b) \$20

30. 5 : 3 : 4

31. (a) 3 : 5   (b) 11 : 5

32. 9 cm^2

33. (a) 296 cm^2 (b) 68 cm

34. 28 cm

Unit 4 - Percentage

**Part 1: Part of a Whole as a Percentage
    (pp. 47-52)**

   60%

1. (a) 26%   (b) 40%

2. (a) 6%   (b) 20%   (c) 6%
    (d) 16%   (e) 9%   (f) 25%

3. (a) 48%   (b) 64%   (c) 40%
    (d) 30%   (e) 40%   (f) 6%

5. 12.5%

6. (a) 25%   (b) 20%   (c) 75%
    (d) 80%   (e) 37.5%   (f) 70%

7. (a) $\dfrac{1}{20}$ (b) $\dfrac{2}{25}$ (c) $\dfrac{1}{2}$
    (d) $\dfrac{3}{20}$ (e) $\dfrac{11}{25}$ (f) $\dfrac{39}{50}$

8. 80%

9. 7.5 %

10. (a) 10%   (b) 90%
     (c) 1%   (d) 3%
     (e) 75%   (f) 0.1%
     (g) 4.5%   (h) 22.5%

11. (a) 0.03   (b) 0.35
     (c) 0.4   (d) 0.86

12. (a) 70%   (b) 30%

13. (a) 55%   (b) \$54

14. \$44

15. 161

16. \$618

Copyright © 2006 SingaporeMath.com Inc., Oregon

Practice 4A (p. 53)

1. (a) $\dfrac{2}{25}$ (b) $\dfrac{1}{4}$

   (c) $\dfrac{1}{2}$ (d) $\dfrac{33}{50}$

2. (a) 0.09   (b) 0.9
   (c) 0.15   (d) 0.62
3. (a) 40%   (b) 87.5%
   (c) 45%   (d) 5%
   (e) 50%   (f) 8%
   (g) 15%   (h) 24.5%
4. 15%
5. 35%
6. 9%
7. 40%
8. 25%
9. 20%
10. 28%

Practice 4B (p. 54)

1. (a) 11.25   (b) 702   (c) $75
   (d) 180 m   (e) 10.5 ℓ   (f) 9.6 kg
2. $60
3. 9
4. 45
5. $190
6. $1470
7. 48
8. 250 g
9. $34
10. 16

Part 2: One Quantity as a Percentage of Another (pp. 55-59)

1. 10%
2. 15%
3. 150%
4. 75%
5. (b) 20%
6. 7.5%
7. 10; 25%
8. $67.20
9. 4.25 kg

Practice 4C (p. 60)

1. 32%
2. 25%
3. (a) 150%   (b) 50%
4. 26%
5. 25%
6. 20%
7. 25%
8. 15%
9. 40%
10. 40%

Part 3: Solving Percentage Problems by Unitary Method (pp. 61-66)

   $3,000
1. 56
2. $960
3. $20
4. 150
5. $1320
6. $750; selling price $900
7. $40
8. 200
9. 90
10. 16

Practice 4D (p. 67)

1. $1200
2. 1100
3. 45
4. 800
5. $200
6. $60
7. $400
8. 80
9. $50
10. 84

Practice 4E (p. 68)

1. 40%
2. 24
3. $1875
4. $50
5. (a) 5   (b) 12.5%
6. 50%

Copyright © 2006 SingaporeMath.com Inc., Oregon

7. 180

8. 16%

9. $2200

10. (a) 32%   (b) $37.50

Review C (pp. 69-73)

1. (a) $\frac{3}{4}$ (b) $\frac{7}{10}$

2. $\frac{1}{5}$

3. (a) $\frac{3}{8}$ (b) $\frac{7}{8}$

    (c) 7   (d) 2

4. $\frac{3}{5}$

5. $\frac{1}{12}$

6. (a) 0.35   (b) 2.2

7. (a) $\frac{3}{40}$ (b) $1\frac{1}{25}$

8. $\frac{3}{5}$

9. 6.96

10. (a) $\frac{1}{6}$ (b) $\frac{1}{4}$

11. (a) 22   (b) 20

12. 17.93 km

13. 0.89

14. (a) 55,000   (b) 4   (c) 4.1

15. 62.5%

16. 23.5%

17. (a) $\frac{4}{5}$ (b) $\frac{1}{20}$

18. (a) 0.06   (b) 0.92

19. (a) 1.5 kg   (b) 210 m

20. 25%

21. 35

22. 64

23. 3900

24. $5.70

25. 25

26. $5.20

27. 300

28. 21 kg

29. $68.40

30. $1400

31. 26

32. $91.80

33. 700 ml

34. 50

35. $\frac{3}{16}$

36. 60 ℓ

37. 3 : 5 : 4

38. $\frac{9}{17}$

39. $9

40. 102

41. 144

42. $3.60

43. 25%

44. $1250

45. 16%

46. 13 cm^2

47. 60 cm^2

48. C

49. (a) 16   (b) 40%   (c) 30%

Unit 5 - Speed

Part 1: Speed and Average Speed (pp. 74-80)

225

1. 50 km/h 65 km/h 35 km/h
    ^{US} 90 mi/h ^{3d} 90 km/h
    ^{US} 45 mi/h ^{3d} 90 km/h
    ^{US} 70 mi/h ^{3d} 90 km/h

2. 100

3. 200

4. 210

5. 5

6. 55

7. 89

8. $3\frac{1}{2}$

9. 8

10. (b) 11:00

11. 5

12. 49

13. 370, 5 , 74

14. 132, 2 , 66

15. 360, 5 , 72

16. 48, 192, 3, 1, 48

Copyright © 2006 SingaporeMath.com Inc., Oregon

Practice 5A (p. 81)

1. 1.25 m/s
2. 30 cm/s
3. ^{US} 1260 mi ^{3d} 1260 km
4. 50 s
5. 11:00 a.m.
6. 52.5 m/min
7. $1\frac{1}{2}$ h
8. 200 m/min
9. 120 km
10. 9 min

Practice 5B (p. 82)

1. 54 km/h
2. (a) 20 km   (b) 1 h 40 min
3. 8 km/h
4. 630 m
5. 2 h 15 min
6. (a) 150 km   (b) 50 km/h
7. (a) 360 km   (b) 60 km/h

Review D (pp. 83-86)

1. (a) thirty thousand, six hundred
   (b) two million, four hundred seventy thousand
2. (a) 600   (b) 0.007
3. (a) 26,327   (b) 43,469
4. (a) 400   (b) 0.005
5. (a) 38   (b) 200
6. (a) 8.7   (b) 2.40
7. 33
8. (a) 36   (b) $16\frac{1}{3}$
9. (a) 10 (b) 1000
   (c) 100 (d) 10
10. 7
11. (a) 3.604   (b) 0.055
12. 3.42
13. 120 ml
14. (a) 1250   (b) 100
15. (a) 1   (b) 36; 27
16. (a) 37.5%   (b) 45%
17. (a) $\frac{21}{250}$ (b) $\frac{7}{25}$

18. 70%
19. (a) 135   (b) 1440
20. (a) 5 : 1 : 4   (b) $\frac{1}{2}$ (c) 50%
21. 4 : 3
22. 60%
23. $14.40
24. 10
25. $0.90
26. $11.75
27. $60
28. 80
29. 21
30. 100
31. 54 cm^2
31. 7 : 5
33. $270
34. $70
35. 500 m
36. 11:10 a.m.
37. (a) 50 km/h   (b) 50 min
38. (a) B   (b) 6 : 7   (c) 30%

Review E (pp. 92)

1. (a) 10,000   (b) 0.4
2. (a) 9   (b) 4
3. 4.501 km
4. 14.5
5. $\frac{1}{4}$
6. (a) 5   (b) 1000
7. $\frac{1}{6}$
8. (a) 1 h 20 min   (b) 2 ℓ 670 ml
9. (a) 70   (b) 800
   (c) 2300   (d) 90
10. 2.63
11. 2.6 kg
12. 64%
13. 2.5%
14. 36
15. $3.24
16. (a) 4 : 5   (b) $\frac{5}{4}$
17. 3 : 4
18. 20

Copyright © 2006 SingaporeMath.com Inc., Oregon

19. 360 g
20. $8
21. $3.55
22. 900
23. 35 g
24. (a) 3   (b) $28,800
25. $7\frac{1}{5}$ m
26. 15 cm
27. 20
28. $1560
29. 0.8 kg
30. $50
31. 15
32. (a) 3 : 4 : 2   (b) $90
33. 50
34. 30%
35. 25%
36. 58%
37. 414
38. 9:00 a.m.
39. 50 km/h
40. 54 cm²
42. 36
43. (a) 3000   (b) 2500   (c) $1600

US Review F (pp. 93-96)

1. (a) 1 lb 10 oz   (b) 2 lb 6 oz
   (c) 1 lb 1 oz
2. (a) 69 lb   (b) 92 lb
3. 4 ft 2 in.

4. $3\frac{1}{4}$ lb; 21 in.; $2\frac{1}{2}$ gal; $\frac{1}{3}$ yd
5. (a) 2 yd 2 ft   (b) 2 lb 4 oz
   (c) 6 gal 2 qt   (d) 0 ft 10 in.
6. 206 yd 2 ft
7. $\frac{1}{4}$
8. (a) 68   (b) 96   (c) 88
9. 2 c
10. Travis, 7 oz
11. 15 : 6 : 16
12. 20 gal
13. (a) 7 lb 12 oz   (b) 2 ft 10 in.
    (c) 3 qt 2 c
14. 25%
15. 50%
17. 317.5 mi
18. 150 in.²
19. 27
20. 122 in.
21. 45 min
22. 1 lb 11 oz
23. $\frac{5}{7}$
24. 595 mi
25. 85%
26. 3 c
27. 243 in.²

Copyright © 2006 SingaporeMath.com Inc., Oregon

Workbook Answer Key

Exercise 1

1. (a) $(m + 2)$kg   (b) 6 kg   (c) 8 kg
2. (a) $\$(x - 5)$ (b) \$6   (c) \$10
3. (a) $\$3n$ (b) \$24   (c) \$30
4. (a) $\frac{w}{4}$ cm   (b) 148 cm
   (c) 152 cm
5. (a) 22   (b) 5
   (c) 45   (d) 20
   (e) 3   (f) 12
   (g) 5   (h) 75
   (i) 18   (j) $\frac{1}{3}$

Exercise 2

1. (a) $\$(6x + 5)$ (b) \$17   (c) \$23
2. (a) $\$\frac{40 - y}{6}$ (b) \$5   (c) \$6.50
3. (a) 10   (b) 3
   (c) 2   (d) 6
   (e) 8   (f) 5
   (g) 40   (h) 14
   (i) 116   (j) 128

Exercise 3

1. (a) $3x$ (b) $4y$
   (c) $5n$ (d) $6p$
   (e) $3x$ (f) $4y$
   (g) $11p$ (h) $2e$
   (i) $4a$ (j) $6k$
2. (a) $2n + 4$ (b) $5a + 3$
   (c) $2 + 9x$ (d) $2a + 5$
   (e) $4d + 2$ (f) $6f + 9$
   (g) $2h + 12$ (h) $6a + 1$
   (i) $2k + 5$ (j) $5x + 5$

Exercise 4

2. (a) 6   (b) 5
   (c) 5   (d) 4

Exercise 5

1. A, C, F, G, H
2. B, C, E, G, H

Exercise 6

1. A, C
2. A, D
3. A, D

Exercise 7

1. C
2. C
3. C

Exercise 8

1. (a) $2 : 3$ (b) $2 : 3 : 5$
2. (a) $3 : 5 : 15$ (b) $15 : 23$
3. (a) $3 : 4$ (b) $1 : 2$
   (c) $2 : 3$ (d) $3 : 4 : 6$
4. (a) $4 : 3$ (b) $1 : 3$
   (c) $5 : 15 : 12$ (d) $3 : 2 : 4$

Exercise 9

1. (a) $\frac{5}{6}$ (b) $\frac{6}{11}$ (c) $\frac{6}{5}$
2. (a) $\frac{2}{3}$ (b) $\frac{3}{2}$
3. (a) $\frac{1}{2}$ (b) 2
4. (a) $\frac{7}{5}$ (b) $7 : 5$

Exercise 10

1. 18 cm
2. 15

Exercise 11

1. (a) $6 : 3 : 5$ (b) 30
2. (a) $\frac{4}{3}$ (b) \$420

Exercise 12

1. (a) US 25 cm   3d 15 cm
       US 50 cm   3d 30 cm
       US 75 cm   3d 45 cm

(b)   72 cm   (c)   90 cm

2.   2.8 kg or $2\frac{4}{5}$ kg or 2 kg 800 g

3.   1.5 ℓ or $1\frac{1}{2}$ or 1 ℓ 500 ml

Exercise 13

1.   19 : 21
2.   1 : 2

Exercise 14

1.   36
2. $80
3.   10

Review 1

1.   Any multiple of 30
2.   (a) 50   (b) 66
3.   (a) $\frac{5}{4}, 1\frac{11}{12}, 2\frac{1}{4}, \frac{12}{4}$

    (b) $1\frac{2}{3}, 1\frac{7}{8}, \frac{18}{6}, \frac{45}{8}$

4.   (a) 1375 ml   (b)   4 kg 750 g
5.   (a) 9   (b)   30
    (c) 7   (d)   6
6. $(9 - p)$
7.   41 kg
8.   24
9. $4.20
10. 18 min
11. 0.56 kg
12. 270
13. $\frac{8}{15}$
14. $10
15. $945
16. 8 ℓ
17. 3 : 4
18. (a) 3   (b)   1 : 4
19. (a) $\frac{7}{5}$ (b) $\frac{7}{12}$
20. (a) 3 : 4   (b) $\frac{1}{4}$ (c)   828
21. 800
22. 56 cm

23. 112 cm^2
24. (a) 10 cm   (b)   72 cm^2
25. $8\frac{1}{2}$ cm^2
26. 170 cm^2
27. $2.80
28. $300
29. $3
30. 2.1 kg or 2 kg 100 g

Exercise 15

1.   (a) 25%   (b) 40%   (c) 28%
2.   (a) 75%   (b) 20%   (c) 87.5%
3.   (a) 48%   (b) 60%
    (c) 60%   (d) 20%
    (e) 15%   (f) 62.5%
    (g) 75%   (h) 24%

Exercise 16

1.   (a) $\frac{1}{50}$ (b) $\frac{3}{20}$ (c) $\frac{6}{25}$
    (d) $\frac{9}{20}$ (e) $\frac{3}{5}$ (f) $\frac{37}{50}$
2.   (a) 30%   (b) 8%   (c) 67%
    (d) 0.4%   (e) 2.5%   (f) 38.5%
3.   (a) 0.02   (b) 0.07   (c) 0.1
    (d) 0.8   (e) 0.25   (f) 0.99

Exercise 17

1.   14%
2.   40%
3.   60%
4.   208

Exercise 18

1.   (a) 60%   (b) $72
2.   540

Exercise 19

1.   5500
2. $300
3. $463.50
4.   increase = 4

Copyright © 2006 SingaporeMath.com Inc., Oregon

Exercise 20

1. 40%
2. 50%
3. 15%
4. 125%
5. 400%
6. 125%

Exercise 21

1. (a) $6   (b) 24%
2. (a) $300   (b) 25%
3. 50%
4. 25%

Exercise 22

1. 75%
2. 50%
3. ^{US} 5.1 gal ^{3d} 5.1 kg
4. $684

Exercise 23

1. (a) $18   (b) 150%
2. (a) $1120   (b) 60%

Exercise 24

1. 260
2. $400
3. $45
4. 150

Exercise 25

1. $60
2. $2200
3. Overall decrease = 5

Exercise 26

1. 2000
2. $250
3. 1250
4. 144

Review 2

1. 4,995,000
2. 3.59 kg
3. Any two: 1, 2, 4, 8
4. (a) 7 h 40 min
   (b) 3 h 40 min
5. (a) 100   (b) 10
6. 3.09
7. (a) 8%   (b) 4.6%
8. 21
9. $\frac{3}{8}$
10. 1 : 4
11. 3 : 2
12. (a) $\frac{2}{3}$ (b) 3 : 4
13. (a) 2 : 3   (b) 200 g
14. 25%
15. $60
16. $33. 60
17. 25%
18. $60
19. 2150
20. 200
21. 56%
22. (a) $2.50   (b) $25   (c) $20 h
23. 6 min
24. 40 cm
25. 120 cm^2
26. Answers can vary.

Exercise 27

1. (a) 50 km/h   (b) 75 km/h
   (c) 90 km   (d) $5\frac{1}{2}$ h

Exercise 28

1. 54 km/h
2. 40 m/min
3. 400 km
4. 900 m
5. $\frac{1}{2}$ h or 30 min
6. 3 h

Copyright © 2006 SingaporeMath.com Inc., Oregon

Exercise 29

1. (a) 400 m/min   (b) 2 km
2. (a) 75 km/h   (b) $2\frac{1}{2}$ h
3. 7:43 a.m.
4. 8 min

Exercise 30

1. 5 km
2. 2 h
3. 9.5 km/h
4. (a) 2 h   (b) 70 km/h
5. (a) 198 km   (b) 66 km/h

Exercise 31

1. 50 km/h
2. 7.5 km/h

Review 3

1. 0.001
2. 21 and 28
3. 69.51
4. (a) 1000   (b) 6
5. 65%
6. $240
7. 40
8. $1\frac{3}{5}$ ℓ
9. 1 : 3
10. $\frac{4}{5}$ ℓ
11. 20
12. 15 min
13. $94
14. 105 cm
15. $675
16. 12.5%
17. 15%
18. 5:00 p.m.
19. $1.80
20. 45
21. (a) 60   (b) 3 : 2

22. $30.90
23. 54
24. $336
25. 75 km/h

Review 4

1. $\frac{3}{8}$
2. (a) 12   (b) $\frac{3}{8}$
   (c) $\frac{3}{5}$ (d) 25
3. $\frac{6}{8}$
4. (a) 0.43   (b) 2.44
5. (a) $4\frac{5}{8}$ (b) $1\frac{2}{3}$ (c) $2\frac{1}{2}$
   (d) $\frac{1}{8}$ (e) $\frac{1}{16}$ (f) $\frac{1}{9}$
6. 65%
7. 0.45
8. $\frac{9}{25}$
9. 2.7 km
10. $\frac{1}{10}$
11. $98
12. $270
13. 24
14. 16 : 15 : 9
15. 100
16. 4 km
17. 142.5 cm^2
18. (a) 13   (b) Thursday   (c) 11.8
19. 306 cm^2
20. (a) no   (b) yes
    (c) yes   (d) no
22. $2.90
23. $24
24. $60
25. 12 km/h

Copyright © 2006 SingaporeMath.com Inc., Oregon